1501366640

AF615010

# *THE APPROXIMATIONS HANDBOOK – the art of nearly*

*C. H. J. BEAVEN*

SIGMA PRESS

Wilmslow, United Kingdom

**First published in 1989 by**

Sigma Press 1 South Oak Lane, Wilmslow, Cheshire SK9 6AR, England.

**British Library Cataloguing in Publication Data**

A CIP catalogue record for this book is available from the British Library.

**ISBN:** 1-85058-119-3

**Typesetting and design by**

Sigma Hi-Tech Services Ltd

Printed in Malta by Interprint Limited.

**Cover design by**

Professional Graphics, Warrington, UK

**Distributed by**

John Wiley & Sons Ltd., Baffins Lane, Chichester, West Sussex, England.

**Acknowledgement of copyright names**

Within this book, various trade names and names protected by copyright are mentioned for descriptive purposes. Due acknowledgement is hereby made of all such protection.

# PREFACE

In a long career in industry and the academic world, I have often been surprised at the lack of objective appreciation of the methods of using approximations and approximation techniques in order to reduce the labour of calculation or to make possible estimates which might otherwise be impractable. To be sure, every mathematics and science or engineering course uses approximate methods, but, often, they are presented somewhat apologetically as though they are a poor substitute for the "real" or "right" answer.

As a very new student trying to find the shear modulus of a steel wire using a torsion pendulum, I was admonished by the professor for allowing the pendulum to swing through too large an angle (30°) when the relationship called only for "small swings". Now in this case, the important angle is the angle of shear, which for a long thin wire is a small fraction of the deflection. I could have used an angle of 100° without appreciably affecting the accuracy of the experiment. This illustrates a common lack of understanding of approximations and the part they play in illuminating very complex relationships and thus making possible their use in practical equipment. This experience made me resolve always to test and question such "ex cathedra" pronouncements. It is frequently surprising to find that many otherwise critical people accept similar statements without making any effort to check them.

There is usually no "right" answer, only the best that can be obtained under the circumstances and it is then totally pointless to try to manipulate dubious measurements with highly precise calculations. The other side of the coin must also be considered and terms or digits must not be neglected without a test to find whether they are significant within the working range.

In writing this book I have tried to present a compendium of methods for applying and testing approximations in a number of different fields, but, of course, it is not possible to cover all the requirements and I can only hope that the techniques are general enough to enable the reader to modify them to suit his or her purpose without undue difficulty.

Guesstimation is intended to be taken seriously and should be used to eliminate costly preliminary preparation of multiple proposals when rough calculation can show that some of them are unacceptable in the particular context. I must also admit that, to me at any rate, it has a certain fascination of its own.

SI units have so many advantages and are are now so widespread that there seemed no point in introducing the extra complication of conversion tables to other, more or less obsolescent systems.

Cliff Beaven　　　　Bad Nauheim　　　　1989

# CONTENTS

# LIST OF SYMBOLS

The usual standard symbols such as π are used without further explanation, but the following list shows the meaning of those which may be less common.

1. General symbols

| Symbol | Explanation | Example |
|---|---|---|
| E | times 10 to the power of | 6.3E3 = 6300 |
| ^ | to the power of | 5^2 = 25 |
| * | times | a*b |
| / | divide | c/d |
| ang | angle | |
| Σ | sum of | Σx |
| ρ | density | |
| log() | natural log | |
| log10() | log (base 10) | |
| log2() | log (base 2) | |
| exp() | e^() | |
| L() | Laplace transform | |
| s | Laplace parameter | |

| Symbol | Explanation | |
|---|---|---|
| T | (tera) | E12 |
| G | (giga) | E9 |
| M | (mega) | E6 |
| k | (kilo) | E3 |
| m or mil | (milli) | E-3** |
| μ | (micro) | E-6 |
| n | (nano) | E-9 |
| p | (pico) | E-12 |
| | (femto) | E-15 |
| a | (atto) | E-18 |

** The use of mil is necessary in order to avoid confusion with the abbreviation for metre.

Note: the mathematical operators have the following priority

| | |
|---|---|
| () | brackets (Highest) |
| ^ | To the power of |
| - | Negation |
| * | Multiplication |
| / | Division |
| MOD | Modulus |
| +,– | Addition and subtraction |
| =, <,> | Relational operators |
| NOT, &, OR | Logical operators (Lowest) |

Lower case letters are used for variables (except in BASIC). **ALL** products are shown with the * symbol. That is: the expression anum = rh + dv means that the variable anum is assigned the value of the sum of the two variables rh and dv. rh is **NOT** the product of r and h. The volume of a sphere, for example is: vol = (4/3)*π*(r^3).

Brackets are used where necessary to avoid confusion. To preserve consistency with program useage, multiple roundbrackets are employed when required instead of differing types of parentheses.

# *PSEUDO-CODE DETAILS*

Owing to the lack of standardisation in programming languages, some sort of pseudo-code is necessary to demonstrate algorithms which may be programmed into differing systems. The choice made in this book has been selected because it is considered that it is easily readable and close to ordinary English.

Each algorithm starts with a declaration list of variables and constants (VAR and CONST) of types: integer (INT), floating point (REAL) or character (CHAR). Arrays (ARRAY) may be in any of these types. As far as possible, integers are restricted to names beginning with i, j, k, l, m or n. Names are in lower case letters. Comments are preceded with the name comment and terminated with end comment. (See the DO loop below).

## Statements and Commands

All commands are in uppercase letters. All command blocks are terminated with a final statement. All individual statements, other than commands, are terminated with a semi-colon. After the declaration the algorithm starts with the command BEGIN and finishes with END which must be the last operative statement. For branching algorithms there may be more than one END.

There are two types of loop.

```
comment: the first type of loop is similar to that used
in FORTRAN, but without the label. endcomment.
1.     DO (i = 1; i = n; i = i + 1;)
       statements
       CONTINUE
2.     WHILE ( condition)
          statements
       ENDWHILE
```

Both types of loop may be nested without restriction, but each loop requires a corresponding termination.

```
Conditional statements.

IF ( condition)
     statements
```

```
ELSE
      statements
ENDIF
```

## Operators

All the standard operators such as + , – , * , / , > , < , ≤ , ≥ and = may be used plus &, OR and NOT. The = is used for logical equality as well as for assignment.

## File, disk and screen commands

| To write to | use |
|---|---|
| screen | WRITESCR |
| disk | WRITEDSK #n |
| printer | PRINT |

To open a file use OPEN "filename" #n. To close a file use CLOSE #n

## Routines

Routine names are in upper case letters followed by .PSD

Subroutines are prefixed with SUBROUTINE and the parameters are enclosed within parentheses.

When a subroutine is called in a program the name and the appropriate parameters are given.

# 1

# GENERAL INTRODUCTION

## Introduction

It is often possible by the judicious use of appropriate approximations to obtain satisfactory answers to otherwise intractable problems. It is the aim of this book to present some approximation techniques which have a wide field of application in many different disciplines. Examples are given and tables showing the difference between the approximate and (presumed) true values are included.

The accompanying definition stresses the fact that a useful approximation should give a satisfactory representation and when, as is usually the case, the correct answer cannot be determined, the approximation must be tested by indirect means. In order to make the best use of the methods shown, it is important to regard an approximation as a significant contribution to the solution of a problem and not just as a convenient or unavoidable simplification. This may call for a change in approach to calculations, but it is felt that the benefits gained warrant the extra effort.

Lists of the major symbols used and the main pseudo–code details are given in the pages at the front. It is strongly recommended that these pages should be read first. Some experience with programming techniques is assumed and the pseudo-code lists are intended to provide guidance only. Where physical, chemical or engineering quantities are given, only SI (Systeme International) Units are employed. If other systems are required, then it is recommended that calculations are carried out using SI and converted as a last step. In some cases, where tradition or familiarity is too strong, the procedures may be changed throughout, but it is stressed that this is a hazardous undertaking fraught with many possibilities for serious errors, especially in guesstimation.

In order to avoid difficulties with expressions which can be be misinterpreted the following definitions are used:

An *approximation* to an object is a representation of that object which differs from the (presumed) true characteristic by a sufficiently small amount to allow the use of the representation as a substitute for the actual object.

The word (presumed) in parentheses is used because a reference or target is always necessary, but the actual value may not be calculable, may not be known, or, in particular cases, may not even exist. For example, the exact value of $\pi$ cannot be calculated, the true mean of a population estimated by a sample is not known whereas the median of an even number of data points does not exist.

The definition is very dependent on the meaning given to the expression "satisfactorily small". This implies that the same approximation can be valid in one set of circumstances, but quite invalid in a different set. Part of the purpose of this book is to provide some guide lines for the selection of the appropriate approximation for specific situations. As an illustration of the effect of changing conditions, consider a map based on a standard Mercator's projection. In this type of projection, compass bearings between two points translate correctly. However, for distances an error is introduced which depends on latitude and longitude. Between the latitudes 60° N and 60° S the East – West distance error is less than 10%, but in the North – South direction the error at latitude 60° is 300% even for distances over which the earth may be considered flat.

An *error* (Symbol Er) is defined by:

$$\text{Er} = \text{true value} - \text{obtained value} \qquad 1.01$$

thus Er can be positive or negative.

The *Absolute Error* (Era) is defined to be $|\text{Er}|$. 1.02

The *Relative Error* (Err) is defined by:

$$\text{Err} = \text{Er}/(\text{true value}) \qquad 1.03$$

The *Percentage Error* (Er%) is defined by:

$$\text{Er\%} = |\text{Err}| * 100 \qquad 1.04$$

Errors in data or measurement may be classified into three groups:*(1) Systematic; (2) Random; (3) Mistakes*. Systematic errors arise from the basic characteristics of a system and can be a definite function of the state variables. If the system is known well enough, it is often possible to correct for systematic errors.

Random errors occur without a definite relation to the state variables, but are nevertheless system dependent. For a long series of experiments or tests under fixed conditions, statistical parameters for the distribution of the errors may be derived and, as the number of tests increases, the mean error will usually tend to zero.

Mistakes occur when data is transcribed, transmitted or processed and they have no definite relation to the system. Under normal conditions an untrained human being tends to make between one and three mistakes for every thousand simple operations. An operation is described as an action such as writing one symbol or making one step in a calculation. This rate may be improved by training, but is seldom better than 1 in 10,000.

The rate of occurrence of mistakes in modern computers is so small that it is very difficult to quantify and practically negligible (<1 in 1E14) except for very long (greater than 10 million floating point operations) and sensitive calculations.

Here it should be noted that *bugs* are program mistakes and not generally attributable to the machine. Apart from mistakes, errors in calculations arise from the limitations of finite computing machines

*Rounding Error* (Erd) is caused by the fact that any number can only be implemented to a finite number of digits.

For example: 1/3 = 0.33333333333........... 1.05
but is, in many single precision computations, represented
by: 0.333333 1.06
dropping all the decimal digits to the right of the sixth place.

*Truncation Error* (Etr) arises because numbers derived from infinite series or processes can only be calculated using a finite number of terms. That is, all terms after a given point must be neglected.

As an illustration of this effect consider cos($x$) when $x = 0.05$ radian.

Now $\cos(x) = 1 - x\text{^}2/2! + x\text{^}4/4! - x\text{^}6/6! + \ldots\ldots$ 1.07
taking the first two terms gives

$\cos(0.05) \approx 0.99875$ 1.08
and Etr < 0.0000003

In this case the truncation error is less than the rounding error that would occur using single precision computation so that no useful purpose would be served by taking more terms of the series.

*Accuracy* is a word which has differing meanings depending on the context, but in this book it is only used in a general sense.

The expression *Accurate to*...denotes a limitation of a result as in

$\pi = 3.141459$ accurate to 6 decimal places 1.09
or as in
100/7 = 14.3 accurate to 3 significant figures 1.10

*Precision* is a dimensioned quantity and is defined as the upper bound of the absolute error in the units of the expressed quantity.

Thus the precision of the result in 1.10 is half of one digit in the first decimal place.

The terms "small" and "large" are purely relative and should not be used without qualification even if their significance can be inferred from the context. In many cases an approximation depends on the fact that a function or variable is neglected, but, unless the value of this quantity or, at least its limits, are known, then the validity of the approximation may remain in question.

When used without qualification, the expression:

"$y$ is small compared with $x$"

may, in general, be taken to mean that $y < x/100$, and similarly the expression:

"$z$ is large compared with $x$"

generally means that $z > x*100$.

However, an effort has been made in this book to avoid such statements and, as far as possible, expressions like

"when $y/x < 0.01$"

are used in place of the less precise useage shown above.

It is often difficult to decide on the scale or level of a sensible approximation especially when the answer is the result of an involved series of calculations or processes. Guide lines for the resolution of this type of problem are given in Chapter 2 and the techniques outlined there are then used in the following chapters.

Most of the methods given do not require advanced mathematical knowledge or ability, but, in some cases, a knowledge of elementary calculus and analysis leads to more complete understanding of the techniques. It must be stressed here that each problem is essentially unique and, although a particular method may be applicable, the validity of the resulting outcome should always be checked to ensure that sensible final values are achieved.

At the end of each chapter exercises on the subject matter of that chapter are given. These may incorporate some of the earlier techniques and it is recommended that at least some of the examples are worked before a method is applied to a real problem.

Given that computing systems are so error free and so fast why is it then necessary to use approximations at all? Firstly, many numbers and expressions are derived from infinite series or sequences and can only be represented approximately in a practical system.

Secondly, measurement data are subject to error and exact calculations, even if possible, when carried out on imprecise inputs can lead to serious misinterpretation of the results.

Thirdly, unnecessarily precise operations take time and when complex processes involving many millions of operations are programmed, even the fastest system can benefit from a well designed approximation technique. Although Finite Element Methods can only be briefly be mentioned in a book of this nature, they form one of the most powerful applications of approximation techniques and for further information the reader is referred to the bibliography.

Fourthly, trends and significant variations can often be obscured by lists of multi–digit numbers or very precise curves, whereas simpler, approximate presentations show such effects much more clearly.

Fifthly, but by no means least, the technique of making a rapid, but satisfactorily close, estimate of the solution to a problem under discussion has, many times, prevented hours of pointless argument.

Having given some of the advantages of approximations, *it must be stressed that their* use should be rigorously confined to their regions of applicability. It is by no means always obvious under what conditions an approximation technique fails and doubtful cases should be carefully examined to check validity.

It is also important to be aware of the dangers of being too familiar with a certain type of approximation. Again the map projection provides a useful example. Because of its convenience and simplicity, Mercator's projection is often used for geographical demonstrations dealing with world–wide subjects. As a consequence many people have the impression that land areas north of latitude 55° are very much larger than their true value and although Greenland seems to be bigger than Australia it has, in fact, about one third of the area of the latter country.

Exercise 1 at the end of this chapter is intended to show the effect of calculation using approximate data while exercise 2 demonstrates that, in certain cases an approximation can give as good or better results than a direct calculation. The answer to exercise 2 depends very much on the machine and the operating system, but if single precision algorithms are used the effects are, in general, quite definite. This example is derived from an actual experiment in which a computer simulation of a separation process gave reliable results when the size of input material was greater than 30 μm, but seemed to indicate that the process failed at smaller sizes. When the appropriate approximation was made it became apparent that the effects were due solely to the limitations of the calculation system.

# Presentation of Computer Algorithms

As there are so many Programming Languages in common use, the algorithms are given in Pseudo Code in the text and the corresponding BASIC programs have been collected in Appendix B. The statements and symbols used for the Pseudo Code are given in the pages at the front of the book. As far as possible, a structured program technique has been used for the algorithms, but this tends to make the BASIC translations rather longer than they need to be. However, the increased ease of readability more than compensates for the extra length. Sub–blocks are indented and multi–line IF statements are used in the Pseudo Code which many BASIC dialects do not support. In BASIC, subroutines do not use dummy parameters so that conversions must be made in the main program.

## Introductory Exercises

1. The diameter of the earth is approximately 12700 km. The approximate speed of an aeroplane is 800 km/hr.

   a) What limits, in your opinion, are implied by these statements?

   b) An aeroplane flies one quarter of the circumference of the earth at an approximate average speed of 800 km/hr. Using your answers to part a, determine the greatest possible difference in the time taken to travel the total distance in the extreme cases.

   c) If it costs an extra £5000/hr to keep the aeroplane in the air unnecessarily, calculate the maximum estimation difference caused by the use of imprecise approximations.

2. The output (y) of a size dependent process is given by:

   $y$ = A*(EXP($x$/B) – (1 + $x$/B + (($x$/B)^2)/2))

   where A = 1000 000 (conversion to output unit)

   B = 1000 (B has units: µm)

   $x$ = size in µm ($x$ varies from 1 µm to 50 µm)

   Using the BASIC listing overleaf or your own program, calculate a table showing $y$ as a function of $x$ for steps of 2 µm in $x$ for the following three cases

   a) Using the expression exactly as given above.

b) *y1* = (A/6)*((*x*/B)^3) (first approx.)

c) *y2* = (A/6)*((*x*/B)^3)*(1 + (*x*/B)/4 + ((*x*/B)^2)/20 + ((*x*/B)^3)/120)
(closer approx.)

```
10   REM  BASIC Listing for Chapter 1. Question 2.
20   LPRINT "  COMPARISON OF METHODS OF CALCULATION "
30   LPRINT "  Y = A*(EXP(X/B)-(1+X/B+(X/B)^2/2)) "
40   LPRINT "  FOR A & X POSITIVE THIS IS ALWAYS > 0 AND"
50   LPRINT "  MONOTONIC INCREASING"
60   LPRINT "   Y IS CALCULATED DIRECT FROM THE EXPRESSION "
70   LPRINT "   USING SINGLE PRECISION"
80   LPRINT "   FIRST APPROX. IS Y1 = A*(X/B)^3/6 "
90   LPRINT "   CLOSER APPROX. IS
100  LPRINT "   Y2 = Y1*(1+(X/B)/4+(X/B)^2/20)+(X/B)^3/120)"
110  LPRINT
120  LPRINT:LPRINT "            TABLE FOR X,Y,Y1, AND Y2"
130  LPRINT
140  LPRINT:LPRINT "    A = 1000 000     B = 1000       "
150  LPRINT:LPRINT
160  LPRINT "    X"," Y"," Y1"," Y2"
170  A=1E6 :B=1000
180  FOR X = 0 TO 50 STEP 2
190  U=X/B
200  Y = A*(EXP(U)-(1+U+(U)^2/2))
210  Y1 = A*U*U*U/6 :Y2=Y1*(1+U/4+U*U/20+U*U*U/120)
220  LPRINT USING "#####";X,
230  LPRINT USING "#########.####"; Y,Y1,Y2
240  NEXT L
250  END
```

# 2

# NUMBERS

## 2.01 Discussion

Numbers are both the raw material and the end product of the calculation process and it is therefore helpful to discuss their characteristics before dealing with the applicable approximation techniques.

Mathematically, the set of all real numbers forms a continuum which can conveniently be represented by points on a straight line stretching from zero to infinity in both positive and negative directions. If any section on this representation, say between the numbers 2.0 and 2 + 1/(2^23) is examined more closely as though magnified by a microscope, it is found to be continuous; there being no limit to the number of *separate* values that can exist between these two points. See fig. 2.01.

Note: In these diagrams the points a and b are separated by one least significant digit in the binary floating point representation using a 23 bit mantissa.

Mathematical number-line

Line is infinite in length and continuous

Point a = 2.0; Point b = 2.0 + 1/(2^23);

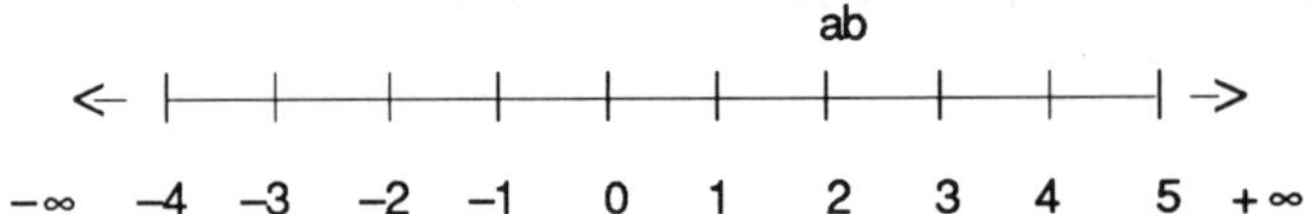


Section near ab magnified by approx. two million times

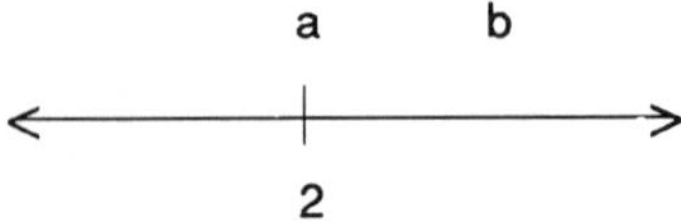


Line is still continuous with infinitely many numbers between a and b.

**Figure 2.01** Mathematical Number–line

This concept is of great use in the theory of numbers, but the situation is altered when practical calculations are undertaken. Then there is a limit both on the largest and on the smallest numbers that can be employed. The number line becomes finite in length and discontinuous as illustrated in fig. 2.02. The length of the line and the gap between each successive number depends on the processor and on the type of program employed. Some typical values are given in Table 2.01 together with the bit structure of the number representation in a commonly used floating point format. Here it should be noted that the same number raised to differing powers of ten does not have similar bit patterns. This fact causes difficulties in rounding and conversion procedures as shown below in section 2.04.6.

Practical Number-line (4 Byte Floating Point)

Line is limited in length and discontinuous

Point a = 2.0;                    Point b = 2.0 + 1/(2^23);

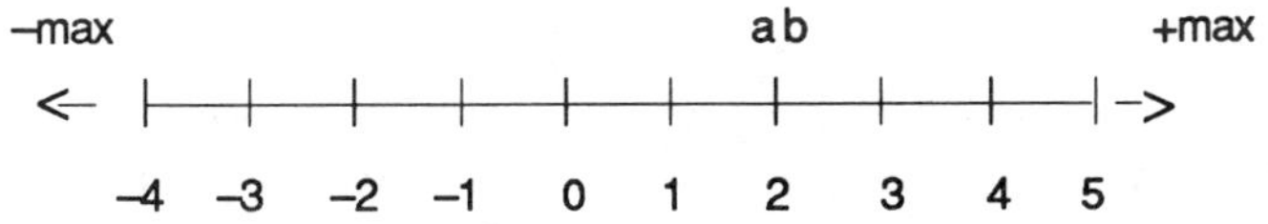

```
 -max                                   ab                   +max
 <-  |----|----|----|----|----|----|----|----|----|->
     -4   -3   -2   -1   0    1    2    3    4    5
```

Section near ab magnified by approx. two million times

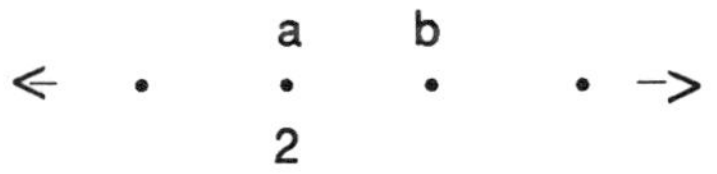

```
            a    b
   <-   •   •    •    •  ->
            2
```

There are no numbers between a and b and the line has empty gaps

**Figure 2.02** Practical Number line.

**Table 2.0.1** Largest and smallest absolute values for various configurations with bit structure for some numbers.

*1 Byte = 8 Bit*

| *No. of Bytes* | *Type* | *Decimal Values Largest/Smallest* | *Least interval* |
|---|---|---|---|
| 2 | Signed Integer | +32767<br>–32768 | 1 |
| | Unsigned Integer | 65535 | 1 |
| 4 | Signed Integer | +2147483647<br>–2147483648 | 1 |
| 4 | Floating Point | 1*10^+37<br>1*10^–37 | 1*10^–6 |
| 8 | Floating Point (Double Precision) | 1*10^+350<br>1*10^–350 | 1*10^–12 |

Note 1: It is possible to decrease the interval in double-precision arithmetic at the expense of range. The actual values depend on the language design and on the processor.

Note 2: For output, the binary numbers must be converted first to Binary Coded Decimal (BCD) and then to string form with suitable rounding. This means that the output precision is worse than that of the actual calculation.

## Typical Bit structure for 4 Byte (32 Bit) Single Precision Floating Point Representations

The last seven bits of Byte 1 plus the left hand bit of Byte 2 form a biassed binary exponent; in this example the bias is 2^128, corresponding to approx. 10^38. On read–out, however, owing to conversion to decimal exponents, the maximum power of 10 is 36.

The left hand bit of Byte 1 is for sign: 1 is –ve. The exponent is adjusted so that the second bit of Byte 2 is always 1. On some systems this bit is removed allowing for one extra bit of precision.

| *Decimal Number* | | | | | *Binary structure* |
|---|---|---|---|---|---|
| | Byte 1 | Byte 2 | Byte 3 | Byte 4 | |
| | < exponent>< | | 23 bit Mantissa | > | |
| 1.0 | 01000000 | 11000000 | 00000000 | 00000000 | 2*(1/2) |
| 2.0 | 01000001 | 01000000 | 00000000 | 00000000 | 4*(1/2) |
| 3.0 | 01000001 | 01100000 | 00000000 | 00000000 | 4*(1/2 + 1/4) |
| 5.0 | 01000001 | 11010000 | 00000000 | 00000000 | 8*(1/2 + 1/8) |
| 500.0 | 01000100 | 11111101 | 00000000 | 00000000 | 512*(125/128) |
| 50.0 | 01000011 | 01100100 | 00000000 | 00000000 | 64*(25/32) |
| 0.5 | 01000000 | 01000000 | 00000000 | 00000000 | 1*(1/2) |
| 0.05 | 00111110 | 01100110 | 01100110 | 01100110 | (1/16)* (3/4 + 3/64 + ..) |
| –20.0 | 11000010 | 11010000 | 00000000 | 00000000 | 32*(1/2 + 1/8) |
| –0.3 | 10111111 | 01001100 | 11001100 | 11001100 | (1/2)* (1/2 + 3/32 + ..) |

The fact that, in most machines, all calculations are done in binary arithmetic introduces a further effect which must be considered in processing numbers.

As mentioned in Chapter 1, a recurring decimal can only be given approximately in a finite system, but fractions such as 1/5 or 7/10 which are terminating fractions in decimal numerals give rise to recurring fractions in the binary system. This effect can produce the irritating consequence that the product 1*10^6*5*10^–6 gives the output 4.99999, whereas the quotient 1*10^6/2*10^5 gives 5.0. In comparisons for jump instructions, therefore, inequalities should be used instead of equalities in order to avoid round–off errors.

In approximation techniques the effect can introduce a bias especially when rounding or converting numbers which arise from values near the system resolution limits.

# 2.02 Number Categories

For convenience, numbers have been divided into several linked categories. The set of all possible numbers that can be used (excluding complex numbers) is called the set of REAL numbers.

NATURAL numbers are positive whole numbers including zero.

INTEGERS include positive and negative whole numbers.

RATIONAL numbers can be expressed as M/N; where M and N are integers

IRRATIONALS cannot be so expressed.

ALGEBRAIC numbers arise as the roots of polynomial equations with integer coefficients. They may be rational or irrational.

TRANSCENDENTALS are real numbers which are not algebraic. They are always irrational.

COMPLEX numbers are stored as two separate real numbers.

VECTORS are arrays of numbers which can have one or more dimensions.

## 2.03 Useful Numbers and their Approximations

Table 2.02 lists some commonly used quantities and numbers rounded to different levels of precision. As it is not possible to include all the values in common use, it is suggested that readers should use a separate sheet on which they can enter the constants which they use frequently together with the relevant accuracy. This helps to keep reference data together in one place and avoids the necessity of time–wasting searches. An additional useful technique is to code the numbers and store them on disk or similar medium so that they may be recalled at will without requiring re–entry. In the C language, for example, if a file named "usflnum.h" is created it can be added to any program with the ' #include' command. Other languages have comparable facilities, but those might require more detailed calling instructions.

Some activities require greater accuracy than others, but it is important to realise that the error given in the table is applicable only to the number itself. The effect of several sequential operations with other approximate values is dealt with in the next section (2.04), but the selection of the appropriate level of accuracy should be made with due regard to the way in which the number is intended to be used. In order to achieve a given minimum error in a final result of a complex calculation, it is, in general, necessary to use primary values with a smaller error.

The table is limited to those approximations intended for practical calculations. Values which are more suitable for "guesstimation" are treated in Chapter 3.

Where the error shown in table 2.02 is significantly greater than the fundamental uncertainty of the value, then the error is signed. If, however, this is not so, then the error entry is preceded by the "±" symbol to indicate that, in order to estimate the resulting error of a computation, the worst case sign should be used. For physical constants the best available value is given, but measurement techniques are being

**Table 2.02** List of useful values with several levels of approximation.

Notes:
1. Suggested storage codes are given after the description
2. As single precision arithmetic rarely uses less than 6 Decimal Digits there is no point in rounding numbers such as **π** to fewer digits except for special purposes (see Chapter 3)
3. In column 5 "E" means "times 10 to the power of"
4. In columns 6 to 10 a "+" means that the true value is greater than the given value

| 1 | 2 | 3 | 4 | 5 | 6 | 7 | 8 | 9 | 10 |
|---|---|---|---|---|---|---|---|---|---|
| Symbol | Name or Description | Store as | Best Value up to 10 Digits | Unit | Error of Best Value (last digit) | Approximations for Errors (in parts per million) less than : 5E-4 | .05 | 5 | 500 |
| e | Nat.Log Base | EXNAT | 2.718281828 | 1 | + | 2.71828183– | 2.718282– | 2.7183– | 2.72– |
| 1/e | e recip. | EXINV | 3.678794412 | E–1 | – | 3.67879441+ | 3.678794+ | 3.6787– | 3.68– |
| π | PI | PI | 3.141592654 | 1 | – | 3.14159265+ | 3.141593– | 3.1416– | 3.14+ |
| 2π | 2*PI | PI2 | 6.283185307 | 1 | + | 6.28318531– | 6.283185+ | 6.2832– | 6.28+ |
| π/2 | PI/2 | PI.5 | 1.570796327 | 1 | – | 1.57079633– | 1.570796+ | 1.5708– | 1.57+ |
| π/4 | PI/4 | PI.25 | 7.853981634 | E–1 | – | 7.85398163+ | 7.853982– | 7.8540– | 7.85+ |
| $\pi^2$ | PI squared | PISQR | 9.869604401 | 1 | + | 9.86960440+ | 9.869604+ | 9.8696+ | 9.87– |
| √π | Sq.rt.PI | RTPI | 1.772453851 | 1 | – | 1.77245385+ | 1.772454– | 1.7725– | 1.77+ |
| 1/√(2π) | Std.Norm. | GAUS | 3.989422804 | E–1 | + | 3.98942280+ | 3.989423– | 3.9894+ | 3.99– |
| √(2π) | Std.Norm.inv | GAUSINV | 2.506628275 | 1 | – | 2.50662828– | 2.506628+ | 2.5066+ | 2.51– |
|  | Deg to Rad | DEGRD | 5.729577951 | E1 deg/rad | + | 5.72957795+ | 5.729578– | 5.7296– | 5.73– |
|  | Rad to Deg | RDDEG | 1.745329252 | E–2 rad/deg | – | 1.74532925+ | 1.745329+ | 1.7453+ | 1.75– |
| N | avogardo No. | AVONO | 6.022169 | E26/kmole | ±6.6 |  |  | 6.0222+ | 6.02+ |
| k | Boltzman | BOLT | 1.380622 | E–23 J/K | ±43 |  |  | 1.3806+ | 1.38+ |
| eV | Electronvolt | ELV | 1.6021917 | E–19 J | ±4.4 |  |  | 1.6022– | 1.60+ |
| c | Light vel | CVEL | 2.9979250 | E8 m/sec | ±0.33 |  | 2.997925± | 2.9979+ | 3.00– |
| $c^2$ |  | CSQR | 8.9875543 | E16 $m^2/sec^2$ | ±0.67 |  | 8.987554+ | 8.9876– | 8.99– |
| h | Planck | PLANC | 6.626196 | E–34 J sec | ±7.6 |  |  | 6.6262– | 6.63– |
| h/2π |  | PLANG | 1.045919 | E–34 J sec | ±7.6 |  |  | 1.0546– | 1.05+ |
|  | Ln π | LOGPI | 1.144729886 | 1 |  | 1.14472989– | 1.144730– | 1.1447+ | 1.14+ |
|  | Ln 10 | LOG10 | 2.302585093 | 1 |  | 2.30258509+ | 2.302585+ | 2.3026– | 2.30+ |
|  | Log10 2 | L10_2 | 3.010299957 | E–1 |  | 3.01029996– | 3.010300– | 3.0103– | 3.01+ |
|  | Log10 π | L10PI | 4.971498727 | E–1 |  | 4.97149873– | 4.971498+ | 4.9715– | 4.97+ |
|  | Log10 e | L10_E | 4.342944819 | E–1 |  | 4.34294482– | 4.342945– | 4.3429+ | 4.34+ |
|  | Log2 10 | L2_10 | 3.321928095 | 1 |  | 3.32192810– | 3.321928+ | 3.3219+ | 4.32+ |

continuously improved and some entries may differ from the current value in the least significant digits.

## 2.04 Calculation Techniques

So that errors and ineffective processes may be avoided, a few remarks on the manner in which most machine operations are carried out have been included, along with the preferred methods.

### 1. Addition and Subtraction

When two floating point numbers are added they must first be converted to the same power range. For example, to add 3.12 to 1.756E1, the number 3.12 is first expressed in the form .312E1 and the sum is then 2.068E1 or 20.68. (Although the example is given in decimal form, the actual calculation will normally be carried out in binary). Therefore, if one number is greater than the other by a factor of more than the reciprocal of the system precision, then the sum will be just the larger number as the smaller number will be ignored. This effect is not usually very important in simple calculations, but it can lead to serious errors in serial additions. In these cases the preferred method is to arrange to add the smaller numbers together before adding the larger ones.

### 2. Multiplication and Division

When a series of numbers are being multiplied or divided, it is good practice to include a test to detect overflow or underflow and to print a warning. This allows action to be taken to prevent a false intermediate result which may give an entirely wrong final answer. The effct of overflow or underflow varies with the operating system and it is therefore, difficult to give more definite guidance. Requirements which are very easy to overlook are that any test number and the result of the test *must* lie within the system limits!

In general, for continued multiplication and division, it is preferred that the intermediate answers are kept near the middle of the system range, where possible. For example, in the calculation of binomial expansions, terms involving quotients of factorials such as $(n!/(n-r)!/r!)*x\char`^r$ may occur. This is best processed as shown in the following pseudo-code fragment:

```
BINTERM.PSD

Comment: calculates binomial term; the values of n,x and
the previous values of i,r and the term (ans) are given.
ans starts with the values ans = n * x; r = 1; and i = n;
endcomment.
```

```
CONST: REAL x; INT: n;
VAR: REAL: ans; INT: i,r;
  BEGIN i = i - 1; r = r + 1;
   IF (i > 0)
       ans = ans * x * i / r;
   ENDIF
 END
```

### 3. Powers

Some languages such as PASCAL and C do not have built–in power functions. This is a deliberate omission because the process used in other languages involves taking logarithms with subsequent reconversion to real numbers. For integer powers, therefore, it is more accurate, and usually quicker, to use a loop as shown in method 2 above. Non–integer powers can simply be implemented by the algorithm:

$$\text{ans} = \exp(\, y * \log(\, x\,))$$

where $y$ is the index and $x$ the number.

### 4. Random Numbers

The fact that the numbers in computers are discontinuous has consequences for the use of the pseudo–random generators provided with most machines. Although the repeating cycle of the generator may be very long, the number of *different* values of the generator output is limited by the length of the memory word used for its storage. Thus, a three byte random number cannot have more than 1677215 different values.

### 5. Many–digit Numbers

Although very high precision calculations may seem out of place in a book on approximations, there are times when the limited accuracy of the machine arithmetic may just not be good enough. Such cases occur, among others, in accounting and surveying. For these purposes, the technique of array arithmetic given in Appendix B can be helpful if the problem concerned does not warrant the purchase or provision of higher accuracy systems. It is written in BASIC, but conversion to other languages is not difficult. The program is slow, but its accuracy is limited only by the maximum dimensions used and the available memory capacity. It is adequate for occasional use.

### 6. Rounding and conversion

The approximate expression of numbers to a precision limited by the practical system used, or by practical requirements, is a very important aspect of calulation procedures. This technique can take several different forms. The rounding methods given here are restricted to decimal numbers, but they can be extended to other systems with relatively simple modifications. (See the examples at the end of this chapter).

There are two commonly used rounding methods: chopping, which simply discards all figures after the chosen limit; and what can be called "normal rounding", which selects that output number which has the required precision and is nearest to the given input.

For measuring equipment, especially for apparatus with digital display or transmission, the output shows a fixed number of digits, with the minimum limited by the system precision. Financial calculations on the other hand must always give their effective output values accurately to the smallest currency unit which can legally be used.

Normally, rounding is used only on print–out for display purposes, but there are cases when it is necessary to round a number during the course of a calculation.

**Example 2.01:**

Consider an investment of £50.00 per month at a yearly interest of 5% compounded monthly over a period of ten years. If this investment is such that it may be withdrawn at the end of any given month, then the interest must be rounded to an exact penny at the end of each month before it is added to the capital.

Table 2.03 shows the different values to be expected at the end of the ten years when the final sum is calculated by four different methods. It may be objected that the differences are trivial and not worth the trouble of making the calculation, but banks and similar institutions make thousands of such transactions every day and over the year the differences can accumulate to a considerable sum. In addition, accountants are expected to balance their books and such differences would be regarded most unfavourably by an auditor.

There is a story, probably apocryphal, about an ingenious programmer working for a bank who added all the fractional pence in his bookkeeping system to a separate storage location and, at the end of each day, transferred the sum to his own account. It is reputed that, at the end of two years, when the ruse was discovered, he had accumulated more than £100,000.

**Table 2.03.** £50.00 invested each month, interest added monthly for 10 years.

| | *Final Sum* | *Calculation Method* |
|---|---|---|
| | £ | |
| 1. | 7846.46 | Using formula (See below) |
| 2. | 7846.49 | Computer (Display rounded to 2 places) |
| 3. | 7846.49 | Computer (Rounded to nearest penny) |
| 4. | 7845.68 | Computer (Rounded down (chopped)) |

Formula. Let: interest be $z\%$ per annum;

sum be final amount;

$r$ be the number of months;

$s$ be the monthly investment.

Then sum = $s * ((1 + z/1200)^{\wedge}(r + 1) - 1) * 1200 / z$

The three computer results were obtained by simple calculation of the monthly interest, adding it to the capital and then adding in the new investment. Where rounding was used, the interest was rounded before adding it to the existing sum.

**Example 2.02:**

The second example shows how rounding may distort an expected result and it often occurs when measurements are digitised before being transmitted. For the purposes of the demonstration the built–in random number generator of the computer was used. This gives a pseudo–random number between 0 and 1. The cumulative sum of many numbers was formed. From each number the value 0.5 was subtracted so that the mean of an infinite sum should be zero if the numbers were uniformly distributed. Table 2.04 shows the results obtained for differing totals using unrounded, rounded to the nearest value and chopped numbers (3 Decimal Places).

**Table 2.04.** Sums of pseudo–random numbers with and without rounding.

| *Total in run* | *Numbers between 0.5 and –0.5* | *Sum* | |
|---|---|---|---|
| | *Not rounded* | *round* | *chopped* |
| 100 | 1.444 | 1.441 | 1.443 |
| 1000 | 1.710 | 1.711 | 1.711 |
| 5000 | 4.700 | 4.693 | 4.719 |
| 10000 | 0.975 | 0.976 | 1.018 |

The differences between the rounded and unrounded values in the above table, although small, illustrate an effect of digitising measured quantities before processing. Integration is often used to reduce noise in signals, but if the signal, including noise, is digitised and then integrated, there may be an extra random component added which is due to the discontinuous nature of the digitiser.

Rounding in the decimal system using a binary machine is a messy procedure as the numbers must first be converted into decimal form, but, as previously mentioned, it is sometimes useful and an algorithm is given.

All machines have built–in conversion routines which may or may not be available. For the latter case and for numbers which can only be expressed in the form "x.xEy" the algorithm given below may be used.

```
CONVERT.PSD (anum,bnum,num(),np,i)

comment: subroutine to convert numbers from binary to
decimal. The result is stored in an array of integers
which can be printed out as required. endcomment.

CONST: REAL t,t1,eta,sysres; INT npt,np,ze;

VAR: REAL anum1,anum,bnum,dv ; INT ko,i,fm1 INT ARRAY
   num(np)

comment: t is the radix of the converted number (10) and
t1 is its reciprocal; np is the largest number of places
required; npt is the value of the stored representation of
the decimal point; anum is the number to be converted
(assumed to be in floating point form); anum1, bnum and
fm1 are dummies for use in the routine; ko and i are
counters and dv is a place shifting multiplier; sysres is
the effective system precision.

If anum has been derived from a decimal number, or is such
that its binary representation is recurring, then, in the
algorithm, it is possible that bnum does not reach anum.

In order to avoid an infinite loop, an inequality is used
in the final WHILE statement.

It is convenient to store the elements of the array num()
in the same format as that used in the machine storage.
Therefore, the integer ze which is equivalent to the
storage value of zero is added to each element as it is
formed. If the number is required for further operations,
ze should be subtracted from each digit.

Alternatively, a separate array, num2() say, can be formed
in the final WHILE loop without ze being added.
CAUTION: bnum is effectively equal to anum, but it is in
machine floating point form and not in BCD. endcomment.

        BEGIN ko = 1;t = 10;t1 = 1/t;anum1 = anum;bnum = 0;
```

```
            dv = 1;i = 1;eta = anum * sysres;
          WHILE (anum1 ≥ t )
             anum1 = anum1 * t1;
             dv = dv * t; ko = ko + 1;
        ENDWHILE
               WHILE ( (anum - bnum) > eta)
                 IF (i = ko + 1)
                    num(i) = npt; i = i + 1;
                 ENDIF
                    fm1 = INT(anum1);
                    num(i) = fm1 + ze;
                    bnum = bnum + fm1 * dv;
                    anum1 = anum1 - fm1;
                    anum1 = anum1 * t;
                    dv = dv * t1; i = i + 1;
               ENDWHILE
                 i = i - 1;
                   IF (ko > i)
                     i = ko;
                   ENDIF
             RETURN (num(),i,bnum)
     END
```

Because of the slightly differing requirements of the ROUND program it was decided to repeat the conversion routine in modified form in the next Pseudo-code example.

```
ROUND.PSD (anum,bnum,l,m,num(arr),typ)

comment: subroutine to round numbers to m decimal places
or l significant digits. The result is returned to the
calling program either in floating point format or as an
array of characters. The variable typ can take the values
c (chop) or r (round). pt is the code for ".". endcomment.

CONST: REAL rrd,t,t1; INT arr; CHAR ze,pt,sig,typ;

VAR: REAL anum,anum1,bnum,dv; INT l,m,ko,i,j,k,nd; CHAR
ARRAY num(arr);

BEGIN
comment: the maximum number of digits (arr) must be
selected first, followed by l,m and typ. If l>0 then m=0
and vice-versa. anum is the signed number to be rounded.
sig is the coded value of the sign. endcomment.

       dv = 1; ko = 1; t = 10; t1 = 0.1;
           rrd = .5; ze = "0"; sig = 0;
           DO ( i = 1; i = arr; i = i + 1;)
               num(i) = ze;
```

```
CONTINUE
        num(0) = 0;
    IF (l > 0 & m > 0)
        WRITESCR "error in data reset l and m";
        INPUT new values for l and m;
    ELSE
        IF ( anum < 0)
           sig = "-";
        ENDIF
           WHILE (anum ≥ t)
          anum = anum * t1;dv = dv * t;ko = ko + 1;
        ENDWHILE
          j = 1; k = j;
        IF ( l > 0)
          nd = l;
        ELSE
          nd = ko + m;
        ENDIF
           WHILE ( k ≤ nd)
           IF ( i = ko + 1)
              num(i) = pt; i = i + 1;
           ENDIF
           IF (( k = nd) & (typ = "r"))
                 anum = anum + rrd;
           ENDIF
           num(i) = INT(anum) + ze;
           bnum = bnum + dv * INT(anum);
           anum = anum - INT(anum);anum = anum * t;
           dv = dv * t1; k =k + 1;
           IF (( l > 0) & ( bnum = 0))
               k = k - 1;
           ENDIF
           i = i + 1;
        ENDWHILE
        i = i - 1; bnum = bnum * sig;
           IF ( ko > i)
               i = ko;
           ENDIF
    RETURN ( bnum,num(),i )
END
```

# 2.06 Justification of Approximations

Calculations made to a precision which is not relevant to the quality of the data used may lead to false conclusions being drawn from the final answers. It is therefore important to know how to present results so that the inherent uncertainty in any value is not only apparent, but can be relied upon. It is much to be regretted that many reports and papers give only the vaguest outline, if any, of the methods used to determine the limits of the output data. Quite frequently, the standard deviation of a number of measurements is used as an estimate of the probable error in the result. This is justified if the deviations are known to be normally distributed, but in systems which are non–linear this is not always so.

This subject is treated in more detail in chapter 4, but some general comments may be relevant in this section.

Where data are derived from measurements with limited precision the actual values obtained should be used in calculations. Other numbers, such as π or e, for example, which are known to much higher precision, should be used to at least two more significant figures than the data. This tends to reduce round-off error in the computation, but care must be taken to see that the machine precision is adequate, especially if complex calculations are being made. For many calculations it can be taken that the total error is not greater than the absolute value of the the sum of the absolute values of the individual errors of the component numbers. (If powers occur, then each component must be counted n times for an index n). In most cases, the error will be much less than this sum because positive and negative errors tend to cancel to some extent.

There are, however two very important cases where this rule does not apply: firstly, where two very nearly equal numbers are subtracted and secondly where one number is divided by a very much smaller one.

**Example 2.03:**

Subtract the two numbers 17.456 and 17.423, each accurate to 5 significant figures; that is, with an error of not more than 0.0005. Their difference is 0.033 with effectively only two significant figures and a maximum error of 0.001, giving a maximum relative error of 1/33.

**Example 2.04:**

Divide 756.32 by .078 – the maximum errors of these two numbers are 0.005 and 0.0005 respectively. The quotient as given by the machine is 9696.41, but if the limiting values – 756.325/.0775 and 756.315/.0785 – are used for the calculation we get 9759.03 as the greatest and 9634.59 as the least answer. In this case we again cannot rely on the result to more than two significant figures and it should be expressed as 9.7E3.

This effect is known as loss of significance and it is well worthwhile to test your data by a few such examples before drawing any conclusions from experiments in which a series of numerical operations has been performed on the output.

If you have any doubt and, perhaps, even when you haven't!, a good rule to remember is:

"Take care of the limits and the places will take care of themselves".

# Exercises 2

In each of the following excercises, single precision should be used. Use your own programming language and system. Results will differ considerably from one machine to another.

1. Using the pseudo–code model CONVERT or otherwise, write programs to convert floating point binary numbers to:

   a). Octal (0 1 2 3 4 5 6 7)

   b). Hexadecimal (0 1 2 3 4 5 6 7 8 9 A B C D E F)

2. Can this be done directly by shifting without using arithmetic ?
   If so, devise a method and use it to express each of the numbers for which the bit structure is given in table 2.01 as:

   a). Octal

   b). Hexadecimal numbers with the appropriate point.

3. Write a program to add the inverse tangent of tan(1 rad) repeatedly (ten times) to an integer starting from 0 and assigning the sum to an integer each time. Print out the result at each stage. Is the answer satisfactory ?

4. What would the value of the constant rrd be if you were rounding in:

   a). Octal

   b). Hexadecimal ?

5. Subtract 175.989986 from 176.0 using only single precision arithmetic and express the result in the form of a six digit number. Hint: express the numbers in the form a + b.

6. Determine the difference between $(LOG(EXP(\sqrt{30000})))^2$ and $(EXP(0.5*LOG(30000)))^2$. Explain the result.

7. Weighing can be a very precise operation and in the most accurate work a correction must be made for the differing buoyancies of the weights and the test article. There are three formulas of which two are exactly equivalent and the third is a close approximation. They are:

   1) $mt = ms * (1 - \rho a/\rho s) + \rho a * Vt$;
   2) $mt = ms * (1 - \rho a/\rho s)/(1 - \rho a/\rho t)$;
   3) $mt = ms * (1 + \rho a * (1/\rho t - 1/\rho s))$; where: $mt$ is the mass of the test piece, $ms$ is the mass of the standard with which it is compared, $Vt$ is the volume of the test piece, $\rho t$ and $\rho s$ are the respective densities and $\rho a$ is the density of air.

   An automatic weighing system weighs blocks of material with a density of 900 kg/cubic metre and nominal dimensions of 10 mm by 50 mm by 110 mm. The resolution of the system is 0.1 mg. The actual volume of the blocks may be taken as constant, but because of process inaccuracies the mass and density can change. Given that $\rho s$ = 8000 and $\rho a$ = 1.3 kg/cubic metre.

   a). Which of the above formulas would you choose to use in the calculation to give the best result ?

   b). If the maximum mass change is 100 mg, what errors are caused if the other formulas are employed without correction for the density change ?

   c). An expensive powder with a density of 2000 kg/cubic metre is distributed throughout the block at a volume concentration of 1/10000. Do the errors in (b) affect the choice of formula for purposes of quality control ?

# 3

# GUESSTIMATION

## 3.01 Definition of Guesstimation

Owing to its nature, a definition of this subject is bound to be rather vague, but the following explanation covers the main features:

*Guesstimation* is the technique of arriving at acceptably approximate estimates of the characteristics of items under discussion using simplified methods and values which enable rapid evaluations and calculations to be made using only mental arithmetic, or, at most, a four-function calculator.

Limiting the calculator to the elementary four-function type may seem unnecessarily restrictive in view of the wide-spread use of more sophisticated, easily portable equipment, but experience has shown that Murphy's law is always with us and it is better to be prepared.

Guesstimation is certainly something of an art form and, like all good art, it is based on a few fundamental rules which enable those of us who are not born to the subject, at least to learn the techniques and make a respectable showing when required. In no sense does guesstimation rely on pure guess–work although when skilfully done the answer may come so rapidly that it appears to be a guess.

The basic rules are given below and they are followed by some explanations which clarify their underlying meaning.

However droll the title may appear, there is no doubt that guesstimation forms an essential part of scientific and engineering discussions, especially when new ideas and projects are being formulated.

Before dealing with the main part of the subject there are two points that should be taken into account.

a) It seems that most people tend to think in linear terms which may cause difficulties when more complex relationships are under consideration.

b) Things are not necessarily what thay appear to be. Often, there may be other reasons for a statement than the purely technical one.

Point (b) will not be further pursued because everybody knows that axes have to be ground. It is, however, important not to be misled by red herrings. Point (a) perhaps needs a little more elaboration. Put the following question to your colleagues:– "I have here four 27 mm ball–bearings. Together they weigh 500 g. Over there is a collection of 9 mm balls which has the same weight. How many are there?" (There are 108)

If the answer is given "off the cuff" you may be surprised at the replies you get.

## 3.02 Rules for Guesstimation

1. Know the reason for and the purpose of the estimate.
2. Know your subject.
3. Identify the region of operation.
4. Identify possible limiting factors.
5. Look for concealed variables.
6. Before opening your mouth and putting your foot in it, check that your answer is reasonable.

Table 3.01 lists some useful numbers and quantities with their relevant guesstimation values.

## 3.03 Explanatory comments

**Rule 1.**
An estimate required for the purpose of comparison between two or more competing proposals, however rough, always needs more care than one for general information. If financial allocations or costs form part of the purpose, then top and bottom limits must be known.

A knowledge of the requirements makes the choice of the accuracy of the estimate possible and appropriate simplifications can be used if applicable.

**Rule 2.**
Do your homework! If you are an expert make sure that uncommon aspects of the

subject cannot catch you out. If not, then, at least, make a study of the parts most likely to be covered.

In this context, two examples from different disciplines may be given. When an operational amplifier is used as part of an analogue computing circuit the actual gain is not important provding it is high enough and the input voltage is neglected compared with the output values. If non–linear components are used for the circuit elements, these assumptions may no longer apply over the same range. Input leakage current and amplifier drift, for example, can both cause significant errors in integrating circuits.

In mineral processing dilute suspensions can be treated as liquids for many purposes, but if there are sharp bends in the pipes through which they are flowing, very severe abrasion may occur.

**Rule 3.**
This can be very important. Many non-linear processes can conveniently be represented piecewise by simpler relationships. Semi-conductor diodes are a case in point. At high currents within their range the voltage-current curve is almost linear with a practically constant potential in series. At lower currents the curve is nearly parabolic and at very small applied voltages the effective resistance of the device is very high.

An extreme example is "bouncing putty". At low shear rates the material behaves like a very viscous liquid, but when an impulse is applied it reacts like an elastic solid.

**Rule 4.**
Simple limiting conditions that come immediately to mind are: freezing and boiling points, the Curie point of magnetic materials, the safe working stress of load bearing structural members and the transition between laminar and turbulent flow (Reynold's number) in fluids.

When a limiting condition occurs in the middle of a chain of processes or reactions, it might be more difficult to spot. Such cases occur in chemical reactions if a product is precipitated, in electronic circuits if an amplifier saturates, in banking when an account becomes empty, in biology when a significant member of a food chain becomes extinct and in engineering when a component fails. If these possibilities are not considered before the estimate is made the result can easily be nonsensical.

In earlier times, the failure of the wind in certain sea areas led to gross errors in the estimate of sailing times to distant parts. For perishable cargoes this was catastrophic.

**Rule 5.**
Concealed variables, from their very nature, cannot always be considered in a guesstimate, but one should be ready to modify a result in the light of better informed opinion. The following example illustrates the sometimes bewildering effect of of a concealed variable.

In a plant extracting a value product from an ore, using a physical–chemical process, it was found that, over a ten year period, the efficiency of the system occasionally dropped drastically in the Spring. All the major variables of the system were controlled including the operating temperature. Although the ambient temperature rose in the Spring, it was eventually discovered that the temperature of the first wash water dropped markedly when the deep source was mixed with melting snow. Consequently, the reaction rate of the first step of the operation was so much reduced by the low temperature that the whole process was affected.

**Rule 6.**
Further comment is really superfluous here except to say that checks should, if possible, be made using a different method from that used for the original result.

# 3.04 How to make a guesstimate

Again a few general rules can be helpful, but practice is necessary in order become proficient. The techniques are illustrated with examples which should be worked through in order to become familiar with the methods.

**3.04.1** Work with not more than 2 significant figures

**3.04.2** Use scientific notation for all values, even pure numbers. That is, express (in your mind) 3000 as 3 k, 0.03 as 30 mil, 0.0003 as 0.3 mil (easier than 300 μ) and so on. The section at the front of the book gives a list of the symbols. (The abbreviation "mil" is used here for 1/1000 to avoid confusion with that for metre).

**3.04.3** Break up every calculation into small steps and simplify the intermediate value before continuing.

**3.04.4** Addition is a lot easier than multiplication, therefore, if you can learn the logarithmic guesstimates in table 3.02, multiple products can be simplified.

**3.04.5** Always carry a "crib–sheet" for important data.

**3.04.6** Never attempt to divide by the difference of two nearly equal quantities. In this case try to recast the calculation as shown in chapter 4.

# 3.05 Examples

**Example 1:**

A steel ring of diameter 449 mm is to be shrink–fitted to a fly–wheel of diameter 450 mm. To what temperature should the ring be raised so that it slides smoothly over the flywheel? The linear coefficient of expansion of steel is between 10 and 15 parts per million over a temperature range from 0 – 500° C.

*Answer.*

Step 1. Take the temperature coefficient to be 10 μ, the required increase to be 1 mm and the ring diameter to be 0.45 metre.

Step 2. Fractional increase is 1/(0.45*1 k) ≈ 2.5 mil

Step 3. Required temperature increase 2.5 mil/10 μ ≈ 250° C.

Step 4. Add this to normal temperature, 25° C. say, giving 275° C.

Step 5. Mentally round up for safety:– approx. 300° C.

A more precise calculation allowing for 0.5 mm clearance gives a required increase of 256° C. Note that, in this example, variations in the metal composition and the coarseness of most oven control systems make accurate calculations relatively meaningless.

**Example 2:**

The earth is about 40 km bigger in diameter at the equator than it is at the poles. How much more does an average man weigh at the poles than at the equator?

*Answer.*

Step 1. Look for hidden variables:– centrifugal effects!

Step 2. Arrange the formulas:

Outward force = $M*w^2*r$.

Inward force = $KM/r^2$, where *K*,*M* and *w* may be treated as constant. *M* is the mass of the man, *K* is a gravitational constant and *w* is angular velocity.

Step 3. Limiting conditions:– at the poles the outward force is zero

Step 4. Let the polar radius be $R$, then the equatorial radius is $R + 20 = R(1 + 20k/R)$ where $R$ is in metres.

Step 5. The inward force at the poles is $KM/R/R$ and at the equator it is $KM/R^2/(1+20k/R)^2 \approx KM(1-40k/R)/R^2$

Step 6. The radius of the earth is approximately 6400 km (6.4 Mm). Therefore the fractional difference in inward force is about 40/6.4 k ≈ 1/0.16 k ≈ 6 mil.

Step 7. The outward force as fraction of the inward force is given approximately by the expression $w^2 * R * R^2/K$. Now $w$, the angular velocity ≈ 2*π/(24*60*60) ≈ 1/(4*60*60) ≈ 1/15 k rad/s.

Step 8. The term $K/R^2$ is equivalent to $g$, the acceleration due to gravity, which is approx. 10 m/s/s.

Step 9. The fraction of step 7 then becomes $1/10*(1/15\ k)^2*6.4\ M$ which is 0.64/0.2 k ≈ 3 mil

Step 10. Therefore the total fractional difference is (careful, they add!) about (3 + 6) mil or say 10 mil ≈ 1%

Step 11. If an average man weighs about 100kg (perhaps a little on the heavy side), then he would weigh an extra 1 kg at the poles compared to his weight at the equator. A more precise calculation for this example gives an increase of 0.872%

# 3.06 Summary

The two examples of section 3.05 illustrate the essential difference in calculation procedures between guesstimation and more accurate techniques. Normally, to achieve accurate results, it is better to make equations from the relevant formulas and then simplify where possible before putting in the numerical values. In guesstimation, however, this procedure can lead to errors because there is usually considerable pressure to arrive at the answer quickly. When each intermediate result is expressed numerically, as demonstrated in the examples above, a constant check can be kept on the validity of the calculation.

# 3.07 Special Types of Guesstimation

Before giving more examples, some methods of estimating separate classes of quantities are presented so that these methods may be included in later calculations. There are a number of general ideas, but, for more specialised disciplines, guide lines

for the development of "local" guesstimators are given. The human-being is a remarkably sensitive observing machine and, with suitable techniques, it is possible to make use of the built–in characteristics to produce useable estimates of external values and quantities with a minimum of extra equipment.

## 3.07.1 Distances

### 1. Horizon distance.

If $R$ is the radius of the earth, $h$ is the height of the observation point and $d$ the horizon distance, then we have:

$$d \approx \sqrt{(2*R*h)} \approx \sqrt{(13*h)}\text{km} \approx 3.6*\sqrt{(h)}\text{ km}$$

where $h$ is in metres.

Therefore, for an eye height of 1.5 m, $d$ is approx. 4.5 km.

At the top of a 30 m rise, $d$ becomes approx. 20 km.

### 2. Distances up to 10 metres.

The distance between the eyes is approximately the same as that across the four fingers of the palm of the hand and the tips of the fingers of a forwardly outstretched arm are about 0.7 metre from the eye. (These values should be verified in your own case). To estimate the distance to an otherwise inaccessible object, raise your right hand and stretch it out in front of you palm upward so that with the right eye shut, the left side of the little finger is lined up with the object. Now shut the left eye and open the right one keeping the arm still. The object will have appeared to have moved to the right to a position within the four fingers. To mark this position, slide the right thumb across the fingers and estimate the distance from the right side of the index finger as a fraction of the distance across the four fingers.

Let this fraction be $f$. Then the object distance $d$ is given by:

$$d \approx 0.7/f \text{ metre}$$

This method can be useful when photographing nearby objects which cannot easily be reached and you have no range–finder.

### 3. Distances up to 100 meters.

The above method can be extended to longer distances if there is some information about the observed object available. For example, most cars are between four and five metres long (in Europe). The angle subtended by the inter–eye distance at arm's length is about 0.1 radian. Therefore if, when the eyes are switched, the palm appears to move a fraction $f1$ of the whole object, then the distance may be estimated from the relation: $d \approx 10*f1*l$ metre, where $l$ is the known length of the object or feature. To give the best results $f1$ should not be less than 1/10 of $l$.

Some standard measurements which are very widely used are shown in Table 3.03

In nearly all cases of distance estimation, the angles subtended by the target objects are such that it can be assumed that the the distance is inversely proportional to the angle.

## 4. Longer distances.

The eye separation base–line is too short for these estimations, so either other reference lengths, or quite different techniques, must be used.

From Table 3.03, it can be seen that the separation of standard gauge railway lines is about 1.5 m and the resolution of the eye is 0.3 milliradian. Therefore the distance at which the lines can just be resolved is approx. 1.5 k/0.3 or 5 km which is similar to the eye–height horizon distance given above.

In this context it is important to note that the apparent transverse separation is inversely proportional to distance and therefore parallel lines receding into the distance appear to be in the form of an hyperbola. This implies that the apparent decrease in longitudinal separation for a fixed increase in distance varies inversely as the square of the distance. If two distant telegraph poles appear to half the distance apart as two nearby ones, assuming the actual distance between them to be 80 metres, then they are about 0.3 km distant; if the longitudinal separation is reduced to one quarter, then they are 1.3 km distant. Street lamp spacing can also be used in this way, but the spacings are not so constant as those for telegraph poles.

The velocity of sound in air at sea–level is about 1/3 km/s. With practice, times of up to 3 or 4 seconds can be estimated to a precision of 0.5 s. If an activity which can be both seen and heard is periodic or pulse–like, the delay between the sight and sound can be used to estimate the distance. We therefore have:– $d \approx$ (delay / 3) km, where the delay is in seconds.

*Please note.* This only works in open situations. In closed areas or within walls or pipes, multiple reflections and wave–guide effects can lead to false results.

## 5. Road distances.

Roads can be conveniently classified as follows:

1. Ordinary roads

2. Motorways
3. Town roads
4. Mountain and river roads

For type 1, the distance between places is up to 20% more than the straight line map distance and for Motorways approximately 5% more. In towns, it is safer to add 30%, but in mountains or near rivers the excess can be as much as 50%. On a long journey, an overall average of between 10% and 20%, depending on the types of road expected, will not often be far wrong. On motorways, it is usually possible to maintain an average speed of about 100 km/hr, but on other roads it is better to assume not more than 50 km/hr. This allows for traffic lights and the occasional tractor. In towns, however, speeds can drop drastically and you may be lucky to achieve 20 km/hr.

## 3.07.2 Velocities

It is a much more difficult to estimate velocity than to judge distance. The problem is complicated by two facts: a) the action is usually transient and b) it takes place in three dimensions. Lines parallel to the line joining the eyes (that is, normally horizontal) seem to be easier to estimate than vertical lines and velocities directly towards or away from the observer have a similar inverse relationship to that given above in 3.07.1.4.

There are basically only two possibilities for directly finding velocities without instruments. The first, which will not be further treated, is the well–known technique of counting the number of objects with a fixed spacing that pass in a known time. The second uses Doppler's principle. Usually, more or less indirect methods must be employed.

**1. Velocities of passing vehicles (for musicians only).**

If a vehicle which is emitting a sound with a recognisable pitch passes close to an observer, the frequency will be observed to drop as the direction changes from approach to recession.

Table 3.04 gives the approximate fractional change of frequency for half–tone intervals on the equal tempered scale and the consequent velocity in km/hour of the moving object.

**2. Relative velocity estimates for vehicles.**

In dangerous or important situations the eye–brain combination is extremely effective in subjectively estimating velocities and in taking the necessary actions. Any attempt to make this process consciously objective is doomed to failure because the mechanism is not externally available and in any case it would be too slow. It is, however, possible to learn to associate known speeds with specific cases as most drivers know, but it requires considerable attention to detail. Therefore, speeds should only be numerically estimated in non–hazardous conditions. In the same manner as for distances, the apparent rate of change of size of an object moving along the line of sight is inversely proportional to the square of its distance. This fact so complicates the judgement problem that, together with the need for a rapid solution, it is not realistic to attempt to give an estimation technique.

**3. Moving liquids.**

On theoretical grounds, the height of the pulse in front of an obstacle in a moving liquid, is proportional to $v^2/g$ and we have:

$$v = \sqrt{(k * g * h)}.$$

In many cases this reduces to: $v \approx \sqrt{(40 * h)}$ where $h$ is in m and $v$ in m/s. (The constant "40" has the units of acceleration.)

Therefore, to estimate the surface velocity of a liquid, it is necessary only to judge the height of the pile produced in front of a convenient obstacle. This, of course, applies also to objects moving relative to the liquid.

In a pipe or open channel, the velocity is not constant throughout the stream and, to determine the volume flow–rate, some additional assumptions are necessary. If the width of the channel is w, then the volume flow rate is given (very approximately) by:

Vol/s $\approx 0.1 * v * w^2$ cubic metre/s

Thus for water flowing in a channel about 2 m wide at a speed of 1 m/s, the mass–flow rate is of the order of 1/2 tonne/s.

## 3.07.3 Areas

For rectangular or nearly rectangular areas with sides greater than 1 m, the side lengths can be estimated as for distances and the area found from the product. For irregular shapes, however, it is usually better to break up the surface into right triangles (not more than three) and estimate the area from these. Sometimes, especially in the case of curved perimeters it is easier to fit circles mentally to the required area. Another method, often applicable to plans, is to imagine a grid superimposed on the surface. The problem is then reduced to counting the number of small grid squares.

It is very important to note here that in a mixture of particles the apparent average area concentration of one particular constituent in a cross section is the same as the average volume concentration for that constituent in the bulk of the mixture. This rather surprising fact forms the basis of many material examination techniques in all branches of engineering and science.

## 3.07.4 Volumes, Masses and Concentrations

If d is the diameter of a sphere, then, with an error of less than 5%, $Vol = 1/2 * d\text{^}3$ or, in words: half the volume of the circumscribing cube. For guesstimation purposes this is quite adequate and much easier to handle than the usual formula. A similar relation holds for ellipsoids of revolution: again the volume is half the volume of the enclosing box.

### 1. Heaps

Many particulate materials, when poured into a heap, tend to form piles having a semi–vertical angle of $\pi/4$ (45°). If the height is h metre, then the geometric volume is approximately $h\text{^}3$ cubic metre. Owing to the large variations in particle shapes and sizes the true volume of the material may be between 50% and 80% of the geometric volume with a typical value of about 60%. Material composed of particles with a large range of sizes tends to pack more closely than material with a uniform size. With the exception of concentrates, most earth–like substances have a density of about 2600 kg/cubic metre and so, for a free standing heap of dry gravel 1 m high the mass would be about 1.6 tonne. There is nearly always water in such a pile and the extra mass amounts to say 300 kg, bringing the total mass to the order of 2 tonne. Where transport costs are concerned, it is very important to compare drying costs with the cost of moving and handling unwanted water.

It is often convenient, for interim storage, to form heaps against a wall. In this case it may be assumed that the volume is then $0.5 * h\text{^}3$. An idea of the value content of such a pile of, say, a concentrate of a copper ore (malachite) 1 m high may be gained as follows: Vol. of pile ≈ 0.5 cubic metre; Vol of ore ≈ 0.3 cubic metre; density of material = 4000 mass of ore ≈ 1.2 tonne; copper content is 58% so that the amount of copper is about 0.7 tonne. At a price of £1400/tonne for copper (March 1988 price) the pile has a value content of about £1000.

The actual value of the pile is less than this amount by the cost of the refining process which can be as much as £100 per tonne.

**Table 3.01** Useful Guesstimation Values.

| *Symbol* | *Item* | *G–Value(s)* | *Symbol* | *Item* | *G–Value(s)* |
|---|---|---|---|---|---|
| $\pi$ | | 3, 22/7 | Vols | Sphere Vol. | 0.5 * (diam)^3 |
| $\pi^2$ | | 10 | | Ellipsoid | $0.5 * (a*b*c)$ |
| | | | | | (a,b,c are maj. axes) |
| e | Nat. log base | 2.7 | | Heap | $h\text{^}3$ |
| 1/e | | 0.37 | | | |
| | $1/\sqrt{(2*\pi)}$ | 0.4 | | | |
| | Deg per rad | 60 | Areas | Circle | $0.8*d^2$ |
| | Rad per deg | 0.02, 1/60 | | Equ.triangle | $0.45*(\text{side})^2$ |

**Table 3.02** Approximate Logarithms.

| *Number* | *log10* | *ln* | *Remarks* |
|---|---|---|---|
| 1 | 0 | 0 | For guesstimation purposes |
| 2 | 0.30 | 0.69 | linear interpolation can be |
| 3 | 0.48 | 1.1 | used between these values. |
| 4 | 0.60 | 1.39 | This applies both to log and |
| 5 | 0.70 | 1.61 | anti–log operations. |
| 6 | 0.76 | 1.79 | It is usually easier, because |
| 7 | 0.85 | 1.95 | of the simple values, to work |
| 8 | 0.90 | 2.08 | in decimals and convert to the |
| 9 | 0.95 | 2.20 | natural system, when required, |
| 10 | 1 | 2.30 | at the end of the calculation. |
| e | 0.43 | 1 | |

**Table 3.03.** Some Standard Values

| *Item* | *Value* |
|---|---|
| Air density | 1.3 kg/cu m |
| Atmospheric pressure | 1E5 Pa |
| Earth radius | 6400 km |
| Light velocity | 3E8 m/s |
| Motorway lane | 3 m |
| Railway gauge | 1.44 m |
| Solar constant | 1.4 kW/m$^2$ |
| Sound velocity | 3.3E2 m/s |
| Telegraph pole separation | 80 m |
| | |
| Standard preferred numbers for | 10 |
| components with 10% tolerance. | 12 |
| The indented numbers are also | 15 |
| uscd for the range with 20% | 18 |
| tolerance. | 22 |
| | 27 |
| | 33 |
| | 39 |
| | 43 |
| | 47 |
| | 56 |
| | 68 |
| | 82 |
| | 100 |

**Table 3.04** Velocities using Doppler's Principle

| *Interval* | *Note* | *Rel freq.* | *%Change* | *Vel. km/hr (6 * %change)* |
|---|---|---|---|---|
| | 1 | 1 | | |
| 1/2 tone | 2 | 1.06 | 6 | 36 |
| Full tone | 3 | 1.12 | 12 | 72 |
| Min. 3rd | 4 | 1.19 | 19 | 115 |
| Maj. 3rd | 5 | 1.26 | 26 | 155 |
| Maj. 4th | 6 | 1.33 | 33 | 200 |
| Aug. 4th | 7 | 1.41 | 41 | At higher speeds the |
| Maj. 5th | 8 | 1.5 | 50 | relation is no longer |
| Min. 6th | 9 | 1.59 | 59 | applicable. For those |
| Maj. 6th | 10 | 1.68 | 68 | with a good ear it is |
| Min. 7th | 11 | 1.78 | 78 | possible to interpolate |
| Maj. 7th | 12 | 1.89 | 89 | between the above values |
| Octave | 13 | 2 | 100 | |

## 3.07.5 Piecewise approximations to curves.

This subject is more comprehensively dealt with in chapter 5, but some guide lines can be given here for rough purposes. Although, in many disciplines, graphs in which both axes are logarithmically scaled are very common, it is strongly recommended that they be avoided if at all possible. They tend to give the impression of nearly linear relationships when this may be very far from the truth. In addition, a log-log plot can obscure other useful approximations, which are more realistic and large abcissa values are cramped together while smaller ones are spread apart. The only valid reason for log–log graphs is a requirement to show a very wide range of values for both variables. Even then, it is better, if possible, to break up the representation into separate sections. Fig. 3.01 shows a growth curve for a population limited by external conditions for a time period of 1000 units. Graph (1) is with linear scales and graph (2) with logarithmic axes. The distortion of graph (2) makes the nearly linear portion between 300 and 600 appear much longer than is actually the case, whereas the initial exponential growth looks much flatter. Regrettably, the portion which appears to be nearly linear, is often projected downwards and upwards to meet the 100% and horizontal axis lines in so–called cut–off points. This reprehensible process is then carried to its logical conclusion and a totally false, simple approximation is derived for the function.

It cannot be too strongly emphasised that this practice is not in accordance with the rules for approximations and it should be not be used.

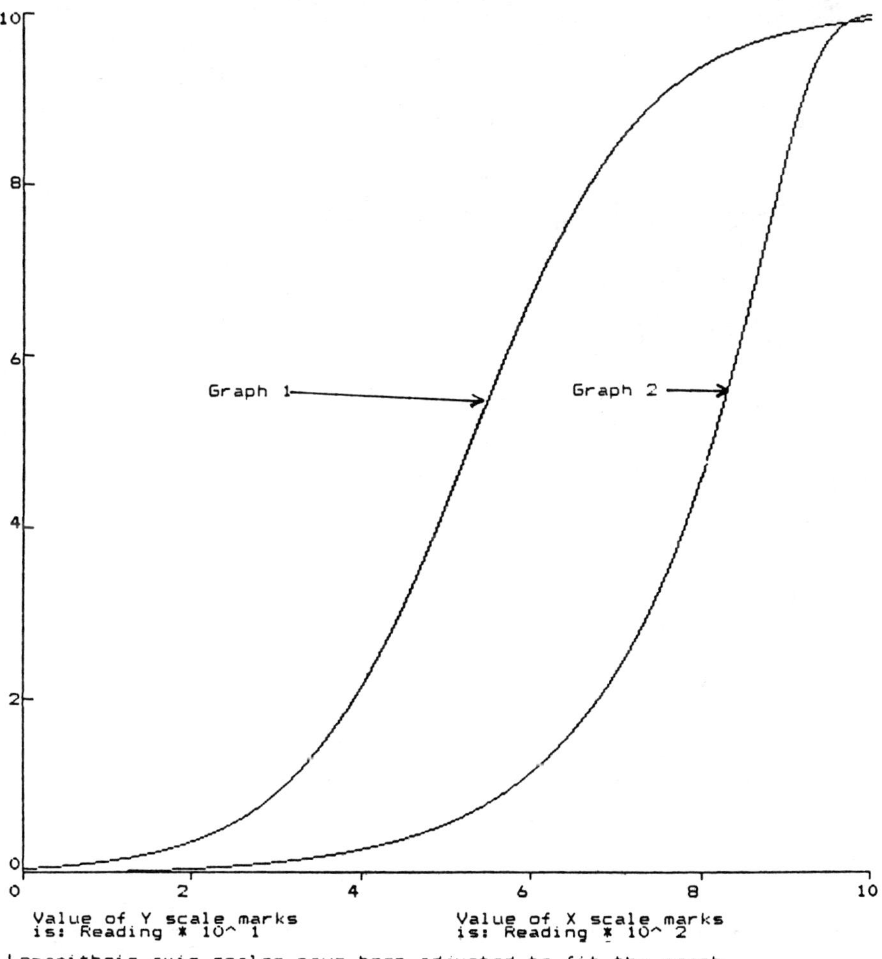


**Figure 3.01** Population Growth: Linear and Logarithmic Scales

For guesstimation, a section of a curve over which the slope changes by up to 30% can be replaced by a straight line. Figs. 3.02–4 show such replacements for a circle, a sine curve and the previously mentioned semiconductor diode curve. The errors of area, length along the curve, value and slope introduced by these replacements are given for each graph in Table 3.05. The diode curve serves to demonstrate some rather important aspects of derived values. The argument of a function such as sin or exp must be a pure number. Therefore, if $V$ represents Volts, the expression exp($V$) is not valid. The form must be changed to exp($c$*$V$), where $c$ has the dimensions of 1/$V$

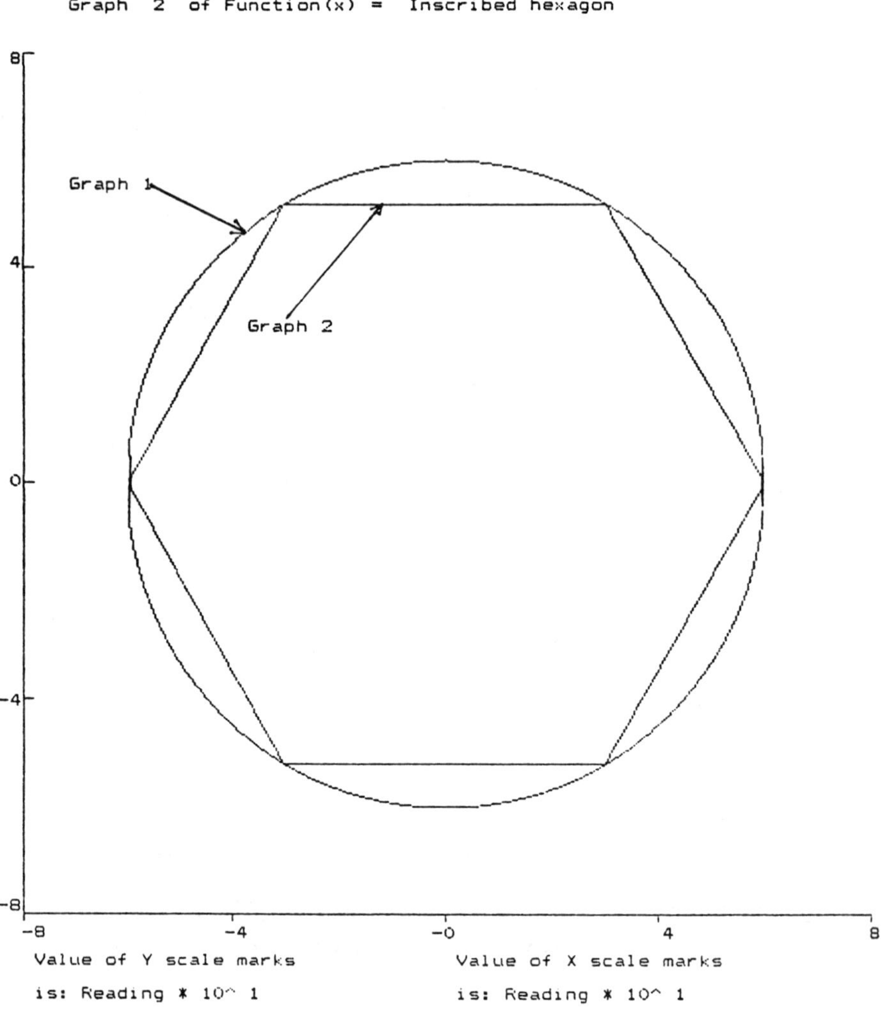


**Figure 3.02** Circle and Linear Approximation

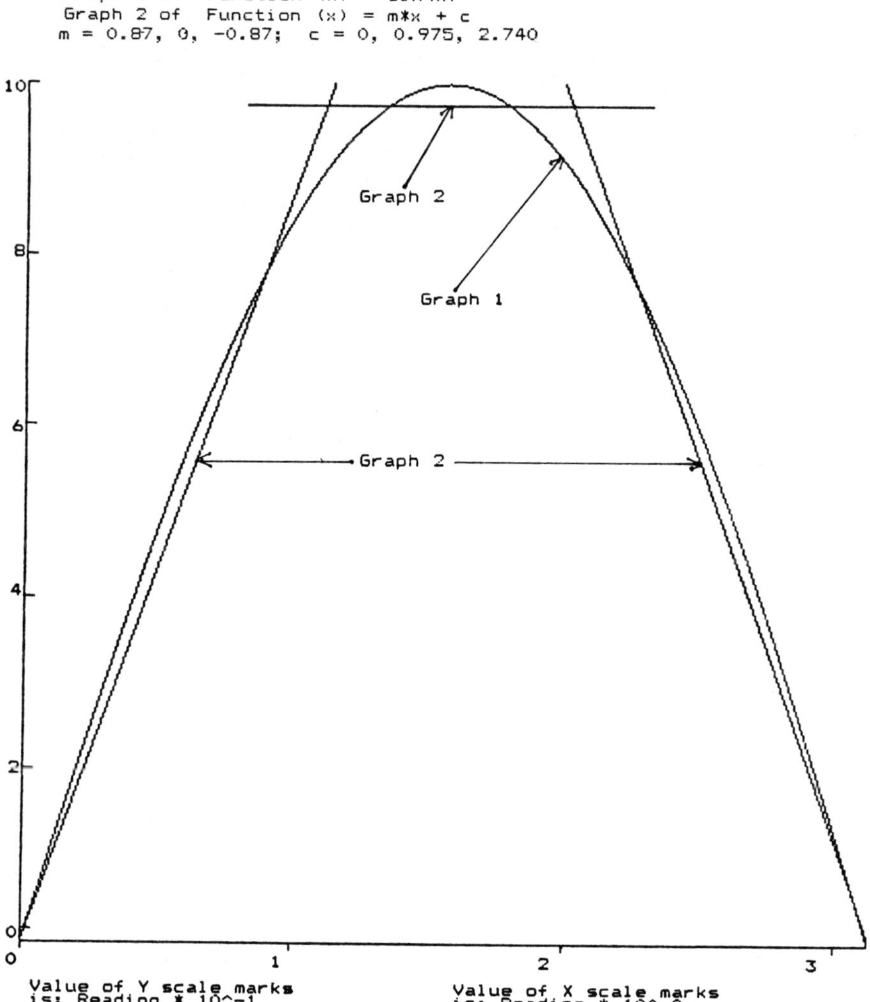


**Figure 3.03** Sine Curve and Linear Approximation

Graph 1 of Function (V) = I (ma/sq mm) = 1E-11*(exp(40*V) - 1)
Graph 2 of Function (V) = I (ma/sq mm) = 1609.35*(V - 0.7)

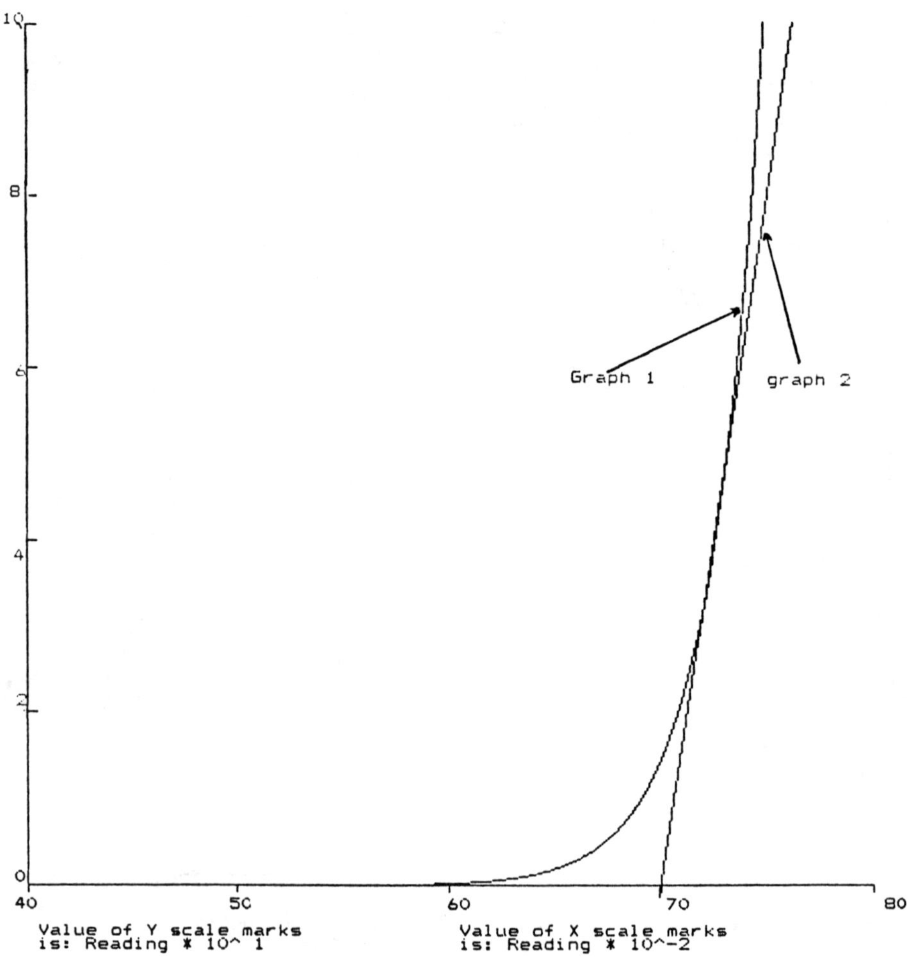


**Figure 3.04** Diode Curve and Linear Approximation

**Table 3.05** Comparison between some curves and their linear approximations

| *Curve* | *Length* | *Area* | *Value* | *Slope* |
|---|---|---|---|---|
| Circle | $2*\pi*r$ | $\pi*r^2$ | $r$ | $-x/y$ |
| Approx. | 6*r | $3*r^2$ | See fig 3.02 | |
| Error | ≈5% | <5% | <13% | up to 26% |
| Sin(x) | 4 | 2 | sin($x$) | cos($x$) |
| Approx. | 3.85 | 1.9 | See fig 3.03 | |
| Error | <4% | <5% | <7% | up to 50% |
| Silicon Diode | $x$ ($x$<1) | ≈ 0 ($x$<1), exp($x$)+$x$–1 | exp($x$)–1 | exp($x$) |
| Approx. | $x$ ($V$<0.7) | Power ($I$*($V$–0.7)) | 0 ($V$<0.7) const*($V$–0.7) | 0 const |
| Error | depends on $V$, | negligible for $V$<0.7 | | |

Note 1: In its direct form the slope of the diode curve gives the incremental conductance which is very small for low values of $V$, but which increases very rapidly when the diode begins to pass an appreciable current. The current is then usually determined by the external circuit except near the break point.

Note 2: The curves demonstrate a general fact that it is easier to integrate using numerical methods than it is to differentiate, although mathematical integration is a more difficult process than differentiation. For example, the area error for the linear approximation to the circle is less than 5%, whereas the slope error can be as much as 26%. Therefore, to attain an error for the slope comparable with that obtained for the area, it is necessary to use a polygon with a much greater number of sides than six.

The basic equation of the diode curve is:

$I = Iz*(\exp(q*V/(k*T)) - 1)$

where

$I$ = current

Iz = a characteristic current of the device

$q$ = electron charge

$V$ = applied potential

$k$ = Boltzmann's constant

$T$ = absolute temperature

At room temperature the term $q*V/(k*T)$ reduces to about $40*V$and $Iz$ is about 1E–14 A for a silicon diode, 1E–6 A for a Germanium diode and 1E–21 A for a LED, With these values, the corresponding curves show a resistance of 1 Ohm at approx. 0.7 V, 0.2 V and 1.1 V respectively. At higher potentials the effective resistance becomes very small, while below these values the devices become effectively non–conductors.

# 3.08 Making your own Guesstimators

Engineering formulas which already contain approximations and assumptions are generally not suitable bases for guesstimation. They often conceal important units in numerical constants and it is therefore safer to start from first principles, if possible.

A useful first step is to derive pairs of values from theoretical or practical cases over the expected range. These should then be plotted on convenient graph paper with a coarse ruling. Ordinary squared paper with a 5 mm grid is usually sufficient. If there are more than two variables, separate sets for the main variables at differing values for each of the subsidiaries should be prepared.

If two independent variables such as temperature and pressure are likely to vary at the same time, a carpet illustrating the three dimensional relationship is sometimes helpful, although it can often be confusing if the user is not thoroughly familiar with the technique.

Having prepared the basic data, the permitted tolerance should be determined. This can be as high as 100% or as low as 5% depending on the application.

Then, suitable model formulas can be tried and compared with the data. Those which give the best results for a minimum of calculation effort should be selected.

# 3.09 General Examples

The exercises at the end of the chapter give a number of problems which can be easily solved with the methods outlined, but, in order to demonstrate the techniques, the following examples are given.

## 3.09.1 Gas Cylinder

A steel gas cylinder has an outside diameter of about 0.25 m, a length of 1.5 m and it can hold oxygen at a pressure of 120 Atmosphere. What is its mass and what mass of gas does it hold? *Data: Density of steel 8000 kg/cubic metre. Gas 1.4 kg/cubic metre. Safe working tensile stress of steel 5E7 Pascal.*

Using the principle of virtual work and neglecting end effects, we find: $S * t = p * (r - t)$; where $t$ = thickness of steel, $r$ = outside radius, $p$ = pressure and $S$ = stress. Therefore,

$t * (S + p) = p * r$; and
$t = p * r / (S + p)$; so,

$t$ = 120E5 * 1/8 / (120 + 500)E5
or $t$ = 12 M * 1/8 / 62 M ≈ 1/40 metre ≈ 2.5 cm

Mass of steel = 3 * 0.25 * 1/40 * 1.5 * 8000 ≈ 200 kg

Mass of gas = 3 * 0.2 * 0.2 * 1.5 * 1.4 / 4 * 120 ≈ 7.5 kg. This mass corresponds to over 5 cubic metres of gas at normal pressure.

## 3.09.2 Tin Market

Which offers the bigger market for tin in Britain, tin–cans for domestic use or domestic electronic equipment? *Data: Population 6E7, thickness of tin–plate 5E–5 m, thickness of flow solder coating 1E–3 m, density of tin 7000 kg/cubic metre.*

Assume the number of households is about 2E7 and each uses 5 cans per week. Approx. average can surface is: 0.02 $m^2$ * 2(sides).

Number of cans per year is 50*1E8 = 5E9 = 5 G.

Therefore mass of tin ≈ 5G * 0.04 * 50 μ * 7 k

= 70 k Tonne. (Cans)

Assume average circuit board size is 0.04 $m^2$

Assume each household buys 2 new pieces of equipment per year.

Amount of solder ≈ 40M*0.08*1 mil*7 k = 20 k Tonne (solder)

Tin content of solder is 40%, so tin use

= 8 k Tonne. (Elec.)

Because of the uncertainties in this calculation the answer is not really definite enough to be relied upon, but the tin–can market is very probably the larger. In this case a knowledgeable expert could give a much closer estimate. No attempt has been made to provide an accurate value, the example is only for the purpose of demonstrating the method.

## 3.09.3 Wind Force

A traffic sign has an area of 6 $m^2$. What is the maximum wind thrust that can be expected? *Data: Max. wind velocity 150km/hr, density of air 1.3 kg/cu. m.* If the wind can flow freely round the sign, the force is given by

$F = \rho * A * v^2 \approx 1.3 * 6 * (150/3.6)^2$

$= 8 * 25 * 25 / 0.36$
$= 14000\ N$ (about 1.4 Tonne force)

### 3.09.4 Geo-stationary Satellite

What is the height of this type of satellite? *Data: Earth radius (r) 6400 km.*

If $R$ is the radius of the orbit, then $R$^3 $= r^2 * g/w^2$

Using logarithms, $\log(R) = (2*(\log(r) - \log(w)) + \log(g))/3$

$\log(R) = (2*\log(6.4M) - \log(40) + 2*(\log(24) + \log(3600)) + \log(10))/3$

$\log(R) = (2*(6.82 + 1.4 + 3.5) - 1.6 + 1)/3$

$\log(R) = (23.44 - 0.6)/3 = 7.61$

$R \approx$ 41000000 m $\approx$ 41000 km

Therefore, the height above the surface is, approximately:

41000 – 6400 $\approx$ 35000 km.

### 3.09.5 Computer Monitor

A computer monitor has a band–width of 18 MHz. How many pixels can be shown on the screen (a) Black and White, (b) Colour? *Data: Assume frame frequency is 60 Hz and there are 600 lines.*

Time of 1 line is 1/(60*600) $\approx$ 25 μs

Rise time of system = 0.4/18 μs $\approx$ 22 ns

Therefore minimum time for 1 pixel $\approx$ 50 ns

Therefore No. of horizontal pixels $\approx$ 25000/50 $\approx$ 500. The screen ratio is 4 * 3, so that the number of pixels is about

500*500*3/4 $\approx$ 180,000 in Black and White and 90,000 in Colour.

The limitation for colour is dependent on the video transfer band–width. If each vertical pixel occupies exactly 2 lines, the number will be reduced to 150 000 (black and white).

# Exercises 3

Each of the following exercises should be worked using the techniques of the chapter. Answers should be limited to two significant figures. The necessary data are either given, can be found in the text, or are well–known.

1. Mentally estimate 10!. (Use logarithms)

2. How many 50 mm diameter solar cells are required to generate 10 kW in full sunlight at:

   a). The equator.

   b). At latitude 50°?

   The efficiency of a new type of solar cell is approx. 20% and the atmosphere absorbs most of the shorter wavelengths.

3. Estimate:

   a). The maximum load of a ready–mixed concrete tanker.

   b). The number of loads required to make a hard standing 50 m * 5 m

   c). If the turn round time of each tanker at the site is half an hour, is it reasonable to deliver all the concrete in one day ?

4. A television station in Los Angeles transmits a signal via a geo–stationary satellite to London.

   a). What is the delay time ?

   b). What is the difference between this time and the telephone delay ?

5. A steel billet 0.3 m square is to be drawn into 2 mm diameter wire in a 5 stage process. The 1st stage rolling speed is 1 m/s.

   a). Is a continuous process posssible ?

   b). If the answer to 5.1 is yes, how many parallel final dies are needed ?

   c). If no, up to which stage should a continuous process be used and what is the diameter of the bar or wire at this stage ?

6. Gold (density 19000 kg/cu. m) occurs at a concentration of one part in 100 000 by weight of the ore in irregular grains about 10 μm in diameter. How many gold grains are there in a lump of ore as big as an orange ?

7. Under average conditions a car requires an output of 15 kW from the engine. How is the fuel consumption affected when all the lights and electrical equipment are switched on ?

# 4

# FUNCTIONAL APPROXIMATIONS: PART 1

## 4.01 Introduction

This and the following chapters deal with the methods of using approximate methods to arrive at problem solutions which give answers that are satisfactorily precise for the application. As explained in the preface, we start from the standpoint that there is no "right answer" where data derived from measurements are involved. No physical law or theory can be "proved" in the sense that some mathematical statements can be justified given certain premises. This is particularly true in the cosmological and nuclear physics fields, where, over the last few years, new theories have replaced the earlier ones with bewildering rapidity, although each successive development brings its own crop of fresh problems.

From this point of view, as it is useless to look for a true value, it is much simpler to begin with the idea that any data processing need only be good enough to achieve the desired end. Formulas can then be simplified, where appropriate, before operating on the data.

In any given problem there may be three types of approximation:

1. Mathematical approximations chosen to simplify calculations.
2. Machine approximations due to the limitations of practical computing equipment.
3. Approximations in the theoretical description of an effect. These may be necessary in order to produce a mathematical model of a process which is amenable to calculation.

These types are not completely independent and the choice of a mathematical approximation is often governed by types 2 and 3.

The rules for appropriate approximations of expressions and formulas may be summarised as follows.

## 4.01.1 Rules for forming approximations

1. Decide on the required precision of the result.
2. Examine the expression and, if possible, identify the largest term.
3. Look for factors and factorise, where possible.
4. Rearrange the expression using brackets so that all constants and mononomials are outside the brackets.
5. Within each bracket (polynomial), take out the largest term as a factor thus setting the largest term to 1. This makes the determination of relative values easier.
6. Where the expression involves sums of terms, each term should, initially, be treated separately and the values compared later.
7. Estimate the relative values of each term and decide whether it may be neglected or not.
8. At this stage the calculation should be examined to determine whether the machine limitations can affect the result. If so, either higher precision must be used or the expression must be rearranged.
9. When possible, calculate both the approximate and the original expression and compare results.

Note: In the case of complicated or indefinite expressions, the original formula may not be calculable and indirect methods of checking must be found as shown in some of the examples.

10. If the comparison in rule 8 is satisfactory, the simplified formula can be used. If not, the procedure should be repeated, retaining the previously neglected terms.

In practice the technique is not so complex as the rules suggest and, frequently simple inspection is enough for a decision, but sometimes, especially in expressions involving quotients or transcendental functions, more thorough examination is necessary. The ideas are developed using frequent examples in order clearly to demonstrate the methods.

Before dealing with engineering and mathematical problems, it is useful to give a programming illustration to show how machine approximations may influence seemingly clear–cut programs.

# 4.02 Loops with non-integer indices

Consider the following Pseudo–code loop.

```
VAR: REAL a,b,c,sum;
    BEGIN a = 0; b = 1.; c = 0.1; sum = 0;
```

```
        DO ( d = a; d = b; d = d + c; )
            sum = sum + d;
            WRITESCR d, sum;
        CONTINUE
    END
```

Now, as explained in chapter 2, 0.1 is a recurring fraction in binary and, as such, it has an error of at least one least significant digit. Thus, when the index *d* is incremented by c an integer number of times (*m*), *d* is either bigger or smaller than $m * c$, depending on the conversion system. If it is bigger, then the final pass will be lost as d becomes greater than *b*. If *d* is smaller than $m * c$, an extra pass through the loop will be made.

In FORTRAN, only integer increments are permitted and this problem does not arise. The effect can be particularly dangerous in nested loops especially if there are jump conditions between the loops. It is strongly recommended that only integer indices and increments be used in order to avoid this problem. The .PSD loop then becomes:

```
CONST: REAL t;
VAR:   INT i,j,k; REAL sum;
    BEGIN t = 0.1; sum = 0; j = 10; k = 1;
        DO ( i = 0; i = j; i = i + k )
            sum = sum + i * t;
            WRITESCR (i * t), sum;
        CONTINUE
    END
```

Because of the integer index the correct number of passes will be made through the loop.

# 4.03 Simple Polynomials

The gas cylinder problem described in Chapter 3, section 3.09, is a suitable first example in which to examine how the successive stages of approximation may be applied in order to arrive at a result that is satisfactorily reliable for practical purposes and this problem is discussed in the following sections.

Consider the expression

$$y = a - b \qquad 4.01$$

where $a > b$ and $a - b$ is much less than $b$.

Now if *a* and *b* are already stored in memory as numbers, there will inevitably be loss of precision as explained in Chapter 2 and there is no possibility of improvement.

If, however, $a$ and $b$ are expressions, the error can sometimes be mitigated by manipulation before evaluation. In estimating the mass of the gas cylinder, the area of a circular ring was approximated by:

$$\text{Area} \approx 2 * \pi * r1 * t \text{ ; where } t \text{ is the thickness} \qquad 4.02$$

$$\text{The correct expression is Area} = \pi * (r1^2 - r2^2) \qquad 4.03$$

Factorising and re–arranging we get:

$$\begin{aligned} \text{Area} &= \pi * (r1 - r2) * (r1 + r2) && 4.04 \\ &= \pi * t * (2*r1 - t) && 4.05 \\ &= 2 * \pi * r1 * t * (1 - t/(2*r1)) && 4.06 \end{aligned}$$

If the ratio $t/(2*r1)$ is small enough, we can neglect it in comparison with 1. This is where the real difficulties begin. What is meant by "small enough" ?

The answer is not very satisfactory and is as follows:

**4.03.1**

A term can be neglected if, when the modified expression is used, the final result may be used for practical purposes without causing dangerous, expensive or misleading effects.

This definition can be used in every case where an approximation technique is employed, but it implies that the decision *must* rest with the user and that, in general, either experience and/or preliminary calculations are necessary before a given simplification can be relied upon.

Table 4.01 shows the error to be expected from the simplified formula (4.02) for $r1$ = 0.125 m (125 mm) and varying values of $r2$. For comparison purposes the calculations for formulas 4.03 and 4.06 are also given in table 4.02 for values of $t$ less than 1 mm.

Examination of Table 4.02 shows that the second formula (4.06) is closer to the double precision result than formula 4.03. This is mainly due to the loss of significance caused by the difference of two nearly equal numbers in formula 4.03. In formula 4.06 the value of $t$ is known in this example and therefore it is the preferred calculation method. However, if we are given $r1$ and $r2$, $t$ is found by subtraction, again leading to a consequent loss of significance. Under these circumstances formula 4.03 is better.

These considerations make it possible to select the best formula for calculations, but do not help with the decision on the level of approximation. For this we must carry the example further.

The annulus area error is only part of the gas cylinder problem and the determination of the stress in the material is important and more difficult. This requirement introduces approximation type 3 which is seldom fully discussed. It refers to the representation of relations between physical magnitudes in the simplest form that gives satisfactory results.

**Table 4.01** Calculation of the area of a circular annulus.

$$\text{area} = \pi * (r1^2 - r2^2) \text{ (Square mm) } t = r1 - r2 \text{ (mm)}$$
$$= \pi * t * 2 * r1 * (1 - t(2*r1)); \quad \text{(double precision)}$$
$$\text{Approx. Area} = \pi * t * 2 * r1; \quad \text{(single precision)}$$

| *t* | *r2* | *Area* | *Approx. Area* | *% error* |
|---|---|---|---|---|
| 0.1 | 124.9 | 7.85084E1 | 7.85398E1 | 0.04 |
| 0.2 | 124.8 | 1.56954E2 | 1.57080E2 | 0.08 |
| 0.4 | 124.6 | 3.13657E2 | 3.14159E2 | 0.16 |
| 0.5 | 124.5 | 3.91914E2 | 3.92699E2 | 0.2 |
| 1 | 124 | 7.82257E2 | 7.85398E2 | 0.4 |
| 2 | 123 | 1.55823E3 | 1.57079E3 | 0.8 |
| 5 | 120 | 3.84845E3 | 3.92699E3 | 2.04 |
| 10 | 115 | 7.53982E3 | 7.85397E3 | 4.17 |
| 20 | 105 | 1.44513E4 | 1.57079E4 | 8.70 |
| 25 | 100 | 1.76715E4 | 1.96349E4 | 11.11 |
| 30 | 95 | 2.07345E4 | 2.35619E4 | 13.64 |
| 40 | 85 | 2.63894E4 | 3.14159E4 | 19.05 |
| 50 | 75 | 3.14159E4 | 3.92699E4 | 25.00 |

In the present case we are concerned with the way in which a material is deformed when a force is applied. The first assumption to be made is that the material is continuous. We know that this is not so, but we try it and compare the result with experimental data. Next, we assume that, for a vanishingly small part of the material, the deformation is a lincar function of the force. With only these two assumptions, the three dimensional deformation is related to the initial size of the part by an array of nine quantities involving the force. This array is called a tensor and such arrays occur frequently throughout science and engineering. Computation using arrays in which each of the nine components are different was, in the past, so time– consuming that further limiting assumptions were introduced in order to make practical calculation possible. Fortunately, most common materials are such that the nine values reduce to two or three if the characteristics are independent of direction. Now that fast computers with large memories are available, many previously intractable problems can be solved.

**Table 4.02** Comparison between different methods of calculation using single precision

| *t* | *Double Precision Area* | *Single Precision Formula 4.03* | *Single Precision Formula 4.06* | *Difference in last 2 places* |
|---|---|---|---|---|
| 0.1 | 7.85084E1 | 7.85060E1 | 7.85083E1 | +23 |
| 0.2 | 1.56954E2 | 1.56951E2 | 1.56954E2 | +03 |
| 0.3 | 2.53337E2 | 2.53340E2 | 2.53336E2 | –04 |
| 0.4 | 3.13657E2 | 3.13659E2 | 3.13656E2 | –03 |
| 0.5 | 3.91914E2 | 3.91913E2 | 3.91913E2 | 00 |
| 0.6 | 4.70108E2 | 4.70106E2 | 4.70107E2 | +01 |
| 0.7 | 5.48239E2 | 5.48235E2 | 5.48239E2 | +04 |
| 0.8 | 6.26308E2 | 6.26308E2 | 6.26307E2 | –01 |
| 0.9 | 7.04313E2 | 7.04313E2 | 7.04313E2 | 00 |

To return to the problem; in Chapter 3 there was an implicit assumption that the radial force was constant throughout the cylinder wall. This approximation made it possible to estimate the tangential stress very easily. Let us now see how far the assumption was justified. At least in the middle part of the cylinder there is cylindrical symmetry, and experience tells us that the vessel tends to expand when filled with gas. In problems with cylindrical symmetry, the relationship between the radius and a dependent variable is often logarithmic and we have as a next step:

$$P = P0 * \log(r/r1) / \log(r1/r2) \qquad 4.07$$

where

$P$ is the pressure at radius r and
$P0$ is the internal pressure

Using this relation the tangential stress in the wall is found to be constant:

$$\text{Stress} = P0 / \log(r1/r2) \qquad 4.08$$

In order to find the wall thickness, we put the stress equal to the safe working value as given previously and solve for $r2$.

$$r2 = r1 * \exp(-P0/\text{stress}) = 0.125 * \exp(-120E5/5E7)$$
$$= 0.0267 \text{ m or } t = 26.7 \text{ mm} \qquad 4.09$$

Comparing this improved value and the area with the chapter 3 results, it can be seen that the thickness is greater by about 6.5%, while the area is smaller than the guesstimation by 12%. The longitudinal stress has not yet been considered and the change in dimensions of the vessel when it is filled has been neglected. Young's modulus for steel is approximately 2E11 $N/m^2$. Thus the change in radius under pressure is about 0.03 mm.

With the above values the longitudinal stress turns out to be 3.8E7 Pa which is within the safe limit and the extension is of the order of 0.5 mm. A gas container has hemi–spherical ends and the stress in each end is the same as the longitudinal stress.

There is now enough information to enable a decision to be made whether the approximation level is good enough. Let us first study the requirements.

1. The cylinder must not explode under pressure.
2. It must not be too heavy to manipulate.
3. It must hold a reasonable amount of gas.
4. It must not be too expensive.

Requirement 1 can never be completely satisfied. The best that can be done is to reduce the chance of failure to an acceptable small value, say 1 in 10 000 000. The so–called safe working stress of steel is based on many tests carried out over a number of years and under different conditions. Even so, each new production batch must be tested individually to see that it complies with the specifications. With the given value of 5E7 Pa there is a safety factor of 5 below the elastic limit. This means that a change of up to 10% in the actual stress will not cause a major problem. Additionally, the stress caused by dropping the vessel from a height of 0.5 m will still be well below the elastic limit.

The stress calculation was based on the unstressed dimensions and is therefore a low level approximation, but as the changes to be expected are less than 0.05%, we can see from the above argument that further refinements are unnecessary in this case.

Calculation of the mass of the container using the value of $t$ in equation 4.09, a cylindrical length of 1.25 m and hemi–spherical ends gives a value of 221 kg. The internal volume is 0.042 cu. m and the mass of oxygen is just over 7 kg.

In this very straight–forward example we have seen how the physical and mathematical approximation techniques are inter– connected and how continual reference back to the end purpose of the application can prevent unnecessarily complex computation. The single precision errors shown in tables 4.01 and 4.02 are, in this case, quite unimportant. Unfortunately, this is not always so and later examples illustrate this problem. Again, the rather close agreement between the guesstimation values and the more accurate calculation is a consequence of the design technique being somewhat over–simplified because of the high safety factor employed. In aircraft design and manufacture, the safety factors must be determined very rigorously to prevent severe weight penalties being incurred.

# 4.04 Binomial expressions

The expression $(a \pm b)$^$p$ 4.10

occurs very frequently in calculations and, in general, it should not be computed directly, but it should be first examined to determine whether applicable approximations can be found. Initially the expression should be rearranged, if necessary, so that the absolute value of a is greater than that of b. If a is negative, the terms within the brackets should be multiplied by –1 and the whole expression by (–1)^p. This operation brings the expression into a form which can then be further arranged to give

$(a$^$p) * (1 \pm b/a)$^$p$ 4.11

Putting

$c = b/a$, the expansion of $(1 + c)$^$p$ for integer $p$ is:

$1 + \Sigma p! * c$^$r/(r! * (p-r)!);$ $r = 1,2, \ldots, p$ 4.12

An algorithm (BINTERM.PSD) for calculating the terms of this expansion for integer $p$ was given in Chapter 2. When $p$ is not an integer, the expression must be expanded in the form of a Maclaurin series (see Chapter 5) and, if $c < 1$, it becomes:

$1 + p*c + p*(p-1)*c$^$2/2! + p*(p-1)*(p-2)*c$^$3/3!$ + etc. 4.13

Table 4.03 shows the maximum errors to be expected for different values of $c$, $p$ and $r$, with $c$ and $p$ less than 1. The remainder $R$ is always less than the figure calculated from:

$R = c$^$(r+1)*(p*(p-1)*\ldots*(p-r-1))/(r+1)!$ 4.14

**Table 4.03** Expansion of $(1 + c)$^$p$ to $r$ terms with maximum remainder $R$

| $c$ | $p = 0.2$ | $R$ % | $p = 0.5$ | $R$ % | $p = 0.8$ | $R$% | $r$ |
|---|---|---|---|---|---|---|---|
| 0.1 | 1.02000 | 0.08 | 1.05000 | 0.12 | 1.08000 | 0.07 | 1 |
| 0.2 | 1.04000 | 0.31 | 1.10000 | 0.45 | 1.16000 | 0.28 | |
| 0.3 | 1.06000 | 0.68 | 1.15000 | 0.98 | 1.24000 | 0.58 | |
| 0.4 | 1.08000 | 1.19 | 1.20000 | 1.67 | 1.32000 | 0.97 | |
| 0.5 | 1.10000 | 1.82 | 1.25000 | 2.50 | 1.40000 | 1.43 | |
| 0.6 | 1.12000 | 2.57 | 1.30000 | 3.46 | 1.48000 | 1.95 | |
| 0.7 | 1.14000 | 3.44 | 1.35000 | 4.54 | 1.56000 | 2.51 | |
| 0.8 | 1.16000 | 4.41 | 1.40000 | 5.71 | 1.64000 | 3.12 | |
| 0.9 | 1.18000 | 5.49 | 1.45000 | 6.98 | 1.72000 | 3.77 | |

| $c$ | $p = 0.2$ | R % | $p = 0.5$ | R % | $p = 0.8$ | R% | $r$ |
|---|---|---|---|---|---|---|---|
| 0.1 | 1.01920 | 0.00 | 1.04875 | 0.01 | 1.07920 | 0.00 | 2 |
| 0.2 | 1.03680 | 0.04 | 1.09500 | 0.05 | 1.15680 | 0.02 | |
| 0.3 | 1.05280 | 0.12 | 1.13875 | 0.15 | 1.23280 | 0.07 | |
| 0.4 | 1.06720 | 0.29 | 1.18000 | 0.34 | 1.30720 | 0.16 | |
| 0.5 | 1.08000 | 0.56 | 1.21875 | 0.64 | 1.38000 | 0.29 | |
| 0.6 | 1.09120 | 0.95 | 1.25500 | 1.08 | 1.45120 | 0.48 | |
| 0.7 | 1.10080 | 1.50 | 1.28875 | 1.66 | 1.52080 | 0.72 | |
| 0.8 | 1.10880 | 2.22 | 1.32000 | 2.42 | 1.58880 | 1.03 | |
| 0.9 | 1.11520 | 3.14 | 1.34875 | 3.38 | 1.65520 | 1.41 | |
| 0.1 | 1.01925 | 0.00 | 1.04881 | 0.00 | 1.07923 | 0.00 | 3 |
| 0.2 | 1.03718 | 0.01 | 1.09550 | 0.01 | 1.15706 | 0.00 | |
| 0.3 | 1.05410 | 0.03 | 1.14044 | 0.03 | 1.23366 | 0.01 | |
| 0.4 | 1.07027 | 0.08 | 1.18400 | 0.08 | 1.30925 | 0.03 | |
| 0.5 | 1.08600 | 0.19 | 1.22656 | 0.20 | 1.38400 | 0.08 | |
| 0.6 | 1.10157 | 0.40 | 1.26850 | 0.40 | 1.45811 | 0.16 | |
| 0.7 | 1.11726 | 0.72 | 1.31019 | 0.72 | 1.53178 | 0.28 | |
| 0.8 | 1.13338 | 1.21 | 1.35200 | 1.18 | 1.60518 | 0.45 | |
| 0.9 | 1.15019 | 1.92 | 1.39431 | 1.84 | 1.67853 | 0.69 | |

Examination of the entries in Table 4.03, where $c$ is between 0.1 and 0.9, shows that the remainder is greatest for $p = 0.5$. Even so, the error falls of rapidly as $r$ is increased. When $c$ is less than 0.1 and $p$ is less than 1, it is often quite satisfactory to terminate the series at $r = 1$. The maximum remainder is then less than 0.125%.

Table 4.03 is intended only for reference; where straightforward numbers are concerned the calculation is best carried out using the machine system if $c$ is greater than 0.01.

# 4.05 Applications

When $c$ is an expression, however, the binomial expansion can be of great help in explaining practical effects.

## 4.05.1 Electronic Integrator circuit

The electronic circuit shown in Fig 4.01 is the basic form of an integrator which is used in many instruments, including digital voltmeters where a precision of 1 digit in six or seven decimal figures may be required. This circuit is well–known, and, for idealised components the transfer relation is given by:

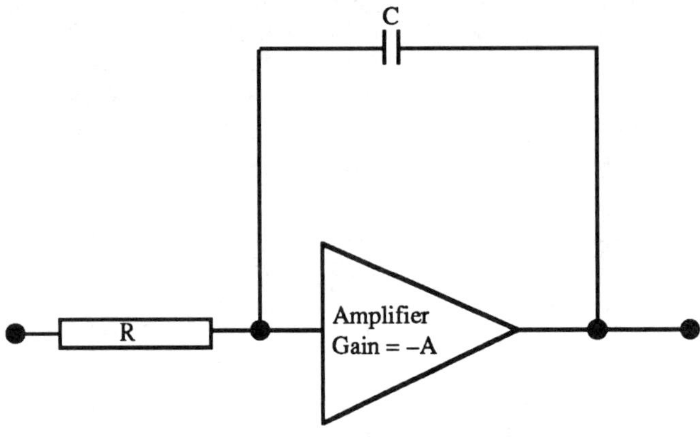


**Figure 4.1** Electronic Integrator Circuit

$$L(vo) = L(-vi*A/((1 + A)*T)*(1/(s + 1/((A + 1)*T)))) \qquad 4.15$$

where $s$ is the Laplace Transform variable. The gain ($A$) of the amplifier is of the order of 1E6 and therefore equation 4.15 is often simplified to:

$$L(vo) = L(-vi\ /\ (s*T)) \qquad 4.16$$

which seems to be a straightforward integral. However, this form gives no indication of how closely the output voltage follows the expected integral. The solution of 4.15 with a step function input is:

$$vo = -A*vi*(1 - \exp(-t/((1 + A)*T))) \qquad 4.17$$

which is very close to 4.16 initially.

To determine the difference between the effect of the simplified form and the more accurate equation (4.17), consider a time constant $t$ of 1 ms, a gain of 1E6 and an input of 1 mV. After a time $t$ = 1s, the output voltage predicted by 4.16 is 1V, whereas 4.17 gives 0.9995 V. This value is not good enough for the type of high quality instrument mentioned above and a more accurate type of integrator design must be used.

If equation 4.15 is written in the form:

$$L(vo) = L(-vi*A/(s*T*(A + 1))*(1/(1 + 1/(1 + 1/s*T*(A + 1))))) \qquad 4.18$$

we can expand the term $(1/(1 + /(1 + 1/s*T*(A + 1))))$ using the binomial relation.

Put $1/(1 + 1/s*T(A + 1)) = c$, then

$$L(vo) = L(-vi*A/(s*T*(A + 1))*(1 - c + c*c - c*c*c + \text{etc.})) \qquad 4.19$$

When the transform of the input is included, each term of 4.19 can be evaluated separately to the required precision. In the present example two terms are sufficient. For a step input, the first term gives the expected linear relation but with a modified coefficient of $A/(T*(A + 1))$ instead of $1/T$. The second term is parabolic with a coefficient of $A/(2*(T*(A + 1)\text{^}2))$. This is about two thousand times smaller than the first coefficient, but, using the expansion it is possible to see very easily what sort of error may be expected, and, if we had used this expansion instead of equation 4.17, the error of 0.0005 V would have immediately been apparent.

This example illustrates several important aspects of the application of approximations.

(a) Standard formulas should not be used without a knowledge of their limitations.

(b) Expressions such as $c$ in 4.19 can be treated as simple variables providing they obey certain conditions which are nearly always fulfilled in practical cases.

(c) With a step function input the solution of 4.19 gives:

$$vo = -A*vi*(t/(T*(A + 1)) - T2/2*(T*(A + 1)^2) + \ldots\text{etc.}) \qquad 4.20$$

which is equivalent to the expansion of 4.17.

(d) Even with the presumption that ideal components had been used, there were still limitations in the circuit. If the leakage resistance and the stray inductance of the capacitor and the input characteristics of the amplifier were included, the deviations from the expected transfer function would become more pronounced. Exercise 4.1 at the end of the chapter deals further with this aspect.

## 4.05.2 Compression of a Van der Waal gas

This example is rather difficult, but it has been chosen to show how an otherwise intractable problem can be solved without complex numerical programs. We will examine the effect of using the Van der Waal gas equation instead of the ideal form in adiabatic compression.

The Van der Waal equation of state for a gas is:

$$(p + a/V^2)*(V - b) = R*T \text{ per mole} \qquad 4.21$$

where: $p$ is pressure; $V$ is volume and $t$ is temperature; $a$, $b$, and $R$ are constants.

The ideal form of 4.21 is

$$p*V = R*T \qquad 4.22$$

Using 4.22 it is relatively easy to derive the adiabatic equation for an ideal gas:

$$p*V\wedge\wp = k \text{ (a constant)} \qquad 4.23$$

but when equation 4.21 is the starting point, the integration cannot be achieved analytically. It must be solved by numerical methods unless an approximations technique is used in the following way:

Although it is not necessary to go through all the steps of the derivation, there are several very useful techniques which can be demonstrated by some of the intermediate working.

From the first law of Thermodynamics and the differentiated form of 4.21, if we assume that the difference between the specific heats $(Cp - Cv) = R$ and that $\wp = Cp/Cv$, we get:

$$dp/p + \wp*dV/(V - b) = (a*dV)*(2/V - 1/(V - b))/(p*V^2) \qquad 4.24$$

Now, the right hand side of 4.24 contains the quotient $dV/(p*V\wedge2)$ which is not directly integrable. However, we expect the solution 4.24 to be not too different from 4.23 if $a$ and $b$ are small and therefore it seems reasonable to replace $p$ by $k/(V\wedge\wp)$ giving:

$$\text{the RHS} = (a*V\wedge\wp)*dV/(V\wedge2*k)*(2/V - 1/(V - b)) \qquad 4.25$$

which can be rearranged as:

$$\text{RHS} = (a*V\wedge\wp)*(1 - 2*b/V)*dV/(k*V\wedge3*(1 - b/V)) \qquad 4.26$$

Expanding $1/(1 - b/V)$ by the binomial technique we get:

$$\text{RHS} = (a*V\wedge\wp)*(1 - 2*(b/V + (b/V)^2 + ..))*dV/(k*V\wedge3) \qquad 4.27$$

which can then be integrated term by term.

Before solving this equation let us examine some relevant data. For a normal modern petrol engine the compression ratio is about 10 : 1. Under working conditions $a = 0.21$, $b = 5.75E-5$, $\wp = 1.3$ and $k = 717$. The volume ($V1$) of one mole at normal temperature and pressure is 0.0224 cu. metre.

These values give: $a/k = 2.93E{-4}$, $b/V1 = 2.57E{-3}$ 4.28

Making the assumption that these values are small enough, the approximate solution of 4.24 becomes:

$$p*(V - b)\hat{}\wp = \text{const}*\exp(-a/(k*(2-\wp)*V\hat{}(2-\wp)) - \ldots) \tag{4.29}$$

giving:

$$p*(V-b)\hat{}\wp \approx \text{const}*\exp(-0.00042/(V\hat{}.7) \tag{4.30}$$

after neglecting the smaller terms and inserting the above values for k and $\wp$.

Thus, to a first approximation, we have:

$$p2/p1 = ((V1/V2)\hat{}\wp)*1.03*\exp(-0.024) = 1.005*(V1/V2)\hat{}\wp \tag{4.31}$$

The effect of $b$ (the volume of the particles) is to increase the compression, whereas the term due to the attraction between the particles ($a$) decreases it. In the example, these two effects almost completely cancel each other showing that, although the mixture is not an ideal gas, the error introduced by assuming the ideal gas laws is less than 0.5%

Recapitulating, the rather complex equation 4.24 was simplified by replacing $p$ and $\wp$ by approximate values and by replacing $1/(1 - V/b)$ by its binomial expansion. Because of the nature of the equation, the justification for the approximation must rest on the fact that difference between the two forms is about 0.5%.

The main advantages gained by the use of the modified formula are: i) the extent of the error incurred by using the ideal form is calculable and ii) the effects of different values of a and b may be estimated.

## 4.05.3 Newton's Law of Cooling

When the temperature of a body is raised in above that of its surroundings it can lose heat by three mechanisms: conduction, convection and radiation. The relevant equations are:

| | | |
|---|---|---|
| Conduction | $dQ1/dt = a1*(T1 - T2)$ | 4.32 |
| Convection | $dQ2/dt = a2*(T2 - T0)\hat{}1.25$ | 4.33 |
| Radiation | $dQ3/dt = a3*(T2\hat{}4 - T0\hat{}4)$ | 4.34 |

where

$Q$ = quantity of heat per unit area
$t$ = time
$T1$ = absolute internal temperature of the body
$T2$ = absolute external temperature of the body
$T0$ = absolute temperature of the surroundings
$a1$,$a2$,a3 are coefficients depending on the configuration of the system.

Newton's law of cooling states that the rate of loss of heat from a body is proportional to the difference in the temperature between the body and its surroundings. We will show that this relationship can be derived from the above equations providing that the temperature difference is not too great and the ambient temperature is taken as approximately 300 K.

Put $\theta = T1 - T2$ and $\Phi = T2 - T0$, ($\theta < 50$°C) then the total rate of loss of heat is

dQ/dt = $(\theta)*a1$ (conduction: inside to outside)

$= a2*(\theta\text{^}1.25)$ (convection) 4.35
$+ a3*\Phi*(T0\text{^}3)*((T2/T0)\text{^}3 + (T2/T0)\text{^}2+(T2/T0)+1))$
(radiation) (outside to surroundings)

Consider the $a3$ term first. This may be expressed as:

$a3*\Phi*(T0\text{^}3)*((1+x)\text{^}3+(1+x)\text{^}2+(1+x) +1)$ 4.36

where $x = \Phi/T0$. ($x < 1/6$)

Then 4.36 becomes

$a3*(T0\text{^}3)*(4+3*x+2*x+x+$ higher terms) 4.37

or, rearranging and neglecting higher powers of x,

$a3*(T0\text{^}3)*4*(1 + w)$, where $w < x*1.5$ 4.38

The $a2$ term is $(a2*\Phi\text{^}1.25) = a2*\Phi * (\Phi\text{^}0.25)$ 4.39

for $T0$ = 300 K, we know that $\Phi < 50$ because $T0 < T2 < T1$.

We may then put $\Phi = 50*(1 - y)$ 4.40

where

$0 < y < 1,$

then 4.39 becomes;

$a2*50*(1 - y)^\wedge 0.25$ 4.41

and, expanding this we get; $50*a2*(1 - y/4+ ...)$ 4.42

The rate of heat transfer from temperature $T2$ to $T0$ must be, by the first law of Thermodynamics, the same as that from $T1$ to $T0$; so collecting the terms and neglecting higher powers, we have;

$a1*\theta = \Phi*(b2*(1 - y/4) + b3*(1 + w));$ 4.43

where $b2$ and $b3$ are new constants. Therefore:

$\theta = \Phi*(b2 + b3)/a1,$ (neglecting $y/4$ and $w$) 4.44

Under most normal circumstances $a1$ is of the order of fifty times as large as $(b2 + b3)$, so that $T2$ is very nearly equal to $T1$ and

$T2 - T0 = \theta + \Phi = \Phi * (1 + ((b2 + b3)/a1))$ 4.45

Then, substituting for $\theta$ in equation 4.35, we find:

$dQ/dt = \Phi*(b2 + b3) = k * (T1 - T0);$ 4.46

(k is a constant depending on the body, the temperature and the nature of the surroundings.)

Equation 4.46 corresponds to Newton's Law of Cooling.

Experimentally, this works quite well for temperature differences of up to 50° C for a wide range of surfaces and conditions.

# 4.06 Factorials

This example shows a method of approximation which can be used when a variable tends to become numerically very large compared with 1. Under these circumstances, if $n$ is the variable, then $1/n \longrightarrow 0$ as $n \longrightarrow \infty$. The expression most commonly used for $n!$ when $n$ becomes large, is Stirling's formula. This can take different forms depending on the application. One form is:

$$n! \approx \sqrt{(2*\pi)} * (n^{\wedge}(n + 0.5))*\exp(-n + 1/(12*n)) \qquad 4.47$$

Table 4.04 gives the values of $n!$, the approximation using Stirling's formula and the consequent error obtained for values of n varying from 5 to 100. Up to 5, $n!$ is manageable. Above 70 the numbers are too big to show in normal form, these are given as logarithms to the base 10. The error entry refers always to the number itself and not to the logarithm.

Equation 4.47 is derived from the asymptotic expansion for the Gamma function:

$$n! = \Gamma\ (n+1) = \sqrt{(2*\pi)} * (n^{\wedge}(n+.5))*\exp(-n)*H \qquad 4.48$$

where $H = 1+1/(12*n) + 1/(288*n^{\wedge}2) - 139/(51840*n^{\wedge}3) - 571/(2488320*n^{\wedge}4) + \ldots$

Putting $x = 1/(12*n)$, the expression for $H$ can be rewritten as:

$$H = 1+x+x^{\wedge}2/2! - 139*(x^{\wedge}3)/(3!*5) - 571*(x^{\wedge}4)/(4!*5) + \ldots \qquad 4.50$$

and

$$\exp(x) = 1 + x+x^{\wedge}2/2!+x^{\wedge}3/3!+x^{\wedge}4/4! + \ldots \qquad 4.51$$

so that

$$\exp(x) - H = ((x^{\wedge}3)/3!)*(1+139/5+(x/4)*(1+ 571/5) +..) \qquad 4.52$$

Therefore $Di$, the difference between 4.47 and 4.48, is given by:

$$Di = \sqrt{(2*\pi)} * (n^{\wedge}(n + 2))\exp(-n)*(26/5)*(1/(12*n))^{\wedge}3 \qquad 4.53$$

and the fractional difference (relative error) by:

$$Err = 265/5*(1/(12*n))^{\wedge}3 \qquad 4.54$$

**Table 4.03** Stirling's Formula and n! for various values of n

| *n* | *n!* 15 Significant Digits | | *Stirling's Formula* 8 Significant Digits | *Error* (ppm) |
|---|---|---|---|---|
| 5 | 120 | | 120.00264 | 22 |
| 10 | 3628800 | | 3628810.05 | 3 |
| 15 | 1307674368000 | | 1.3076754E12 | 1 |
| 20 | 2432902008176640000 | | 2.4329028E18 | .4 |
| 25 | 1.55112100433310E25 | | 1.5511212E25 | .2 |
| 30 | 2.65252859812191E32 | | 2.6525288E32 | .1 |
| 40 | 8.15915283247898E47 | | 8.1591531E47 | .04 |
| 50 | 3.04140932017134E64 | | 3.0414093E64 | .03 |
| 60 | 8.32098711274139E81 | | 8.3209871E81 | .02 |
| 70 | Log10(70!) | = | 100.0784050 | .009 |
| 80 | Log10(80!) | = | 118.8547277 | .006 |
| 90 | Log10(90!) | = | 138.1719358 | .004 |
| 100 | Log10(100!) | = | 157.9700037 | .003 |

The term $\exp(1/(12*n)$ is sometimes omitted from the expression; then we have a simplified Stirling's formula:

$$n! = \sqrt{(2*\pi)} * (n^{\wedge}(n + 0.5))*\exp(-n) \qquad 4.55$$

Table 4.04 gives values for this expression together with the associated error. As n becomes larger the two formulas tend to converge although, as the table shows, the convergence is very slow. As the extra computing effort required for formula 4.47 is so small and the error is mostly less than 1/1000 of the simpler form, the use of formula 4.55 is not recommended.

**Table 4.04** Stirling's Simplified Formula and n! for various values of n

| *n* | *n!* | *Formula 4.55* | *Error (%)* |
|---|---|---|---|
| 5 | 120 | 118.02 | 1.7 |
| 10 | 3628800 | 3598695.62 | 0.83 |
| 15 | 1307674368000 | 1.3004307E12 | 0.55 |
| 20 | 2432902008176640000 | 2.4227868E18 | 0.42 |
| 25 | 1.55112100433310E25 | 1.5459594E25 | 0.33 |
| 30 | 2.65252859812191E32 | 2.6451709E32 | 0.28 |
| 40 | 8.15915283247898E47 | 8.1421726E47 | 0.21 |
| 50 | 3.04140932017134E64 | 3.0363445E64 | 0.17 |

The factorials in tables 4.03 and 4.04 were calculated in full using the array arithmetic program in the Appendix and were rounded to 15 digits for the tables. The time taken to calculate and print out all the factorials from 10 to 60 took about 20 s. 60! has 82 digits.

# 4.07 Limit of $(x^n)/n!$

As a corollary to section 4.06, the important limit of $(x^n)/n!$ as n tends to infinity is easily seen using equation 4.47.

We have, approximately,

$$(x^n)/n! \approx (x^n)/\sqrt{(2*\pi)}*(n^(n + 0.5))*\exp(-n + 1/(12*n)) \quad 4.56$$

If we let $x \longrightarrow \infty$ as $n \longrightarrow \infty$, we get:

$$\text{RHS} \longrightarrow \exp(-n)/\sqrt{(2*\pi)} * (n^(0.5)) \qquad \text{as } n \longrightarrow \infty \quad 4.57$$

and therefore

$$(x^n)/n! \longrightarrow 0 \qquad \text{as } n \longrightarrow \infty \quad 4.58$$

whatever value is given to x.

This result is usually obtained from the limit rule:

$$\text{Lim } (f(x)/g(x)) = d(f(x))/dx/d(g(x))/dx \qquad \text{as } x \longrightarrow a \quad 4.59$$

but the method leading to equation 4.58 gives a much clearer picture of the expression.

If x is less than 1, the value of the LHS of 4.56 converges rapidly to 0, but for $x$ greater than 1, the terms can increase before they eventually decrease. For example: 10^4/4! is 416.67, 10^5/5! is 833.33, but 10^25/25! is 0.64 and 10^30/30! is 0.004.

# 4.08 Rational Functions

Rational functions are expressions of the form:

$$y = f(x)/g(x); \quad 4.60$$

where $f(x)$ and $g(x)$ are polynomials

If $g(x)$ has factors of the form $a*x + b$ it is often tempting to ignore rule 5 in 4.1 1 and expand them directly as $(a*x + b)^(-1)$ using the binomial expansion. It is then quite easy to overlook the possibility that $a*x = -b$ at some value of $x$. If this occurs, y has a discontinuity at $x = -b/a$ and the expansion is certainly not valid. The best procedure is to follow rule 5, then the term becomes a*x*(1 + $b/(a*x)$) and the possibility of $b/(a*x)$ being –1 can easily be examined. Of course, for the expansion to be valid, $b/(a*x)$ must always be less than 1. If $b$ becomes larger than $a*x$, then the rearrangement takes the form:

$$b*(1 + a*x/b)$$

Chapter 5 deals with this subject in more detail.

# 4.09 Polynomial Evaluation

A polynomial consists of the sum of terms like:

$$C(j)*(x^j) \qquad 4.61$$

where

$C(j)$ is an array of constants.

As mentioned in Chapter 2, Section 2.04.3, it is very rarely justified to use the built–in power algorithm for evaluating polynomials. In "Numerical Recipes", Press et al (bib 1) there is a complete and refreshing discussion of this subject in section 5.3, but the folowing pseudo–code fragment illustrates their technique of nested multiplication and addition which drastically reduces the computing load and gives the most accurate answer possible.

```
POLYNOM.PSD(n,coef(n),pol)
comment: n is the highest power of x. The coefficients are
stored in a previously determined array coef(), starting
with 0; pol is the returned value. endcomment.

    CONST: INT n,coef(n) VAR:
    INT k; REAL x,pol;
    BEGIN
       pol = coef(n); k = n-1;
          WHILE ( k ≥ 0)
            pol = pol*x + coef(k); k = k - 1;
          ENDWHILE
       RETURN pol
    END
```

# 4.10 Polynomial Multiplication and Division

For the sake of completeness the algorithms for multiplication and division of polynomials are included here. These are frequently required and, in a slightly modified form, are also very useful for evaluation of complex series to a given precision.

```
SUBROUTINE MULPOL.PSD (n1,n2,coef1(n1),coef2(n2),coef3
(n3))

comment: n1 and n2 are the highest powers of x. The
coefficients are stored in previously determined arrays
coef1()and coef2(), starting with 0. The product
coefficients are returned in coef3() with n3 = n1 + n2.
All the elements of coef3() must have previously been set
to zero. endcomment.

   CONST: INT n1,n2,n3; REAL coef1(n1),coef2(n2);
   VAR:   INT i,j,k; REAL coef3(n3);
   BEGIN
     i = 0;j = 0;k = 0;
           WHILE (k <= n1+n2)
             DO (i = 0; i = n1; i = i+1;)
               DO (j = 0; j = n2; j = j+1;)
                 IF (k = i+j)
                 coef3(k) = coef3(k)+coef1(i)*coef2(j);
               ELSE
                 END IF
                 CONTINUE;
               CONTINUE
             CONTINUE
               k = k+1;
           ENDWHILE
      RETURN coef3();
   END

SUBROUTINE DIVPOL.PSD(n1,n2,cf1(n1),cf2(n2),cf3(n1),rm(n2))

comment: This code segment is based on an algorithm by
Knuth, Bibl(7),section 4.6. n1 is greater than or equal to
n2; that is, cf1() represents the numerator and cf2() the
denominator. The coefficients are stored in previously
determined arrays cf1()and cf2(), starting with 0. The
quotient coefficients are returned in cf3(n1) and the
```

```
remainder coefficients in rm(n2). endcomment
   CONST: INT n1,n2; REAL cf1(n1),cf2(n2);
   VAR: INT i,j,k; REAL cf3(n1),rm(n2);
   BEGIN
     DO (i = 0; i = n1; i = i+1;)
         rm(i) = cf1(i); cf3(i) = 0;
     CONTINUE
         DO (j = n1-n2; j = 0; j = j-1;)
            cf3(j) = rm(n2+j)/cf2(n2);
             DO (k = n2+j-1; k = j; k = k-1;)
                  rm(k) = rm(k) - cf3(j)*cf2(k-j);
             CONTINUE
         CONTINUE
           rm(n2) = 0;
     RETURN cf3(); rm();
   END
```

# Exercises 4

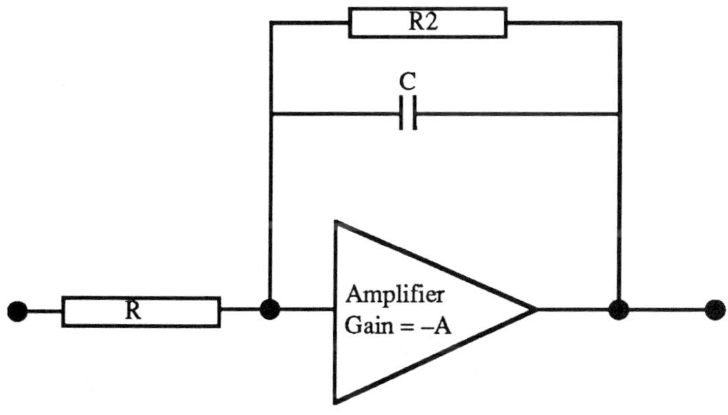


**Figure 4.02** Electronic Integrator Circuit

1. Figure 4.02 shows the diagram of an electronic integrator circuit in which the capacitor (*C*) has a leakage resistance (*r2*). The Laplace transform equation for this circuit is:

   $vo = -vi*A*r2/(r2 + (A + 1)*r1*(1 + s*r2*C))$ 4.62

a) By rearranging the equation and using the binomial expansion, show that, if $r2 = (A+1)*r1$, then the approximate departure from the ideal relation is twice that given in equations 4.19 and 4.20.

b) Discuss how this result affects the choice of components $r1$ and $C$ in practical circuits, given that it is difficult to achieve leakage resistances of greater than 2E11 Ohm without special precautions.

2. An orifice plate flowmeter is used to measure the average flowrate of a liquid with a pulsating flow. The relationship between volume flow ($vm$) and pressure ($p$) is given by:

$$vm = \sqrt{(c*p)}; \qquad 4.63$$

where c is a constant and p, the measured variable, is the pressure drop across the orifice.

Show that, if the flowrate is v1 for one half of the time and v1/2 for the other half, the indicated value given by the flowmeter is nearly 1.5% too great. Assume that the pulses are too rapid to be shown on the indicator. (Use the binomial expansion).

Note: A similar effect occurs in all non–linear measuring equipment when the input is fluctuating and the indicator or output system is calculating the average value.

3. The equation

$$y = (x - 2)/(x - 3)^\wedge 2 \qquad 4.63$$

represents a rational function. Show that:

a) as $x$ becomes greater than 30 or less than –30, $y$ tends to $1/x$

b) by using the substitutions $x = 1 + h$; $x = 2 + h$; and $x = 3 + h$ respectively, where $h < x/10$, the value of $y$ near $x = 1$ is $-0.25*(1 - h^\wedge 2)$, near $x=2$ it is $h*(1 + 2*h)$, and near $x = 3$ it is $1/h^\wedge 2 + 1/h$.

4. Evaluate the binomial coefficient 20!/(9!*11!).

a) Exactly using long integer arithmetic.

b) Using single precision real arithmetic.

c) Using Stirling's formula.

d) Using the simplified Stirling's formula.

Estimate the error in each case.

5. Consider the equation

$$x^2 - 50*x - 0.35 = 0 \qquad 4.64$$

The solution is usually expressed as:

$$x = (50 \pm \sqrt{(2500 + 1.4)})/2 \qquad 4.65$$

a) Show that the solution can be obtained more quickly and more accurately using the binomial expansion with only two terms than by calculation with single precision arithmetic.

b) If 4.64 is written as:

$$x = 50 + 0.35/x \qquad 4.66$$

an immediate approximate solution is obtained by putting x0 = 50

giving

$$x = 50 + 0.007 \qquad 4.67$$

What has happened to the other solution ?

6. Check your interpretation of MULPOL.PSD by modifying it to print out Pascal's triangle for $n = 10$. The entry for $n = 0$ should be in the middle. (Hint: include a counter in the relevant loops.)

# 5

# FUNCTIONAL APPROXIMATIONS: PART 2

## 5.01 Introduction

Most of the functions used in applied mathematics can be expressed in the form of power series or, in some cases, as continued fractions. Continued fractions are mainly used for special functions and will not be further considered here.

Some series can be used for all values of the variable, but usually the series is valid only for a restricted range of values and, quite often, useful only for a more restricted range.

In general, a series used as an equivalent to a given function is useful only if it gives a sufficiently close approximation to the function in less than about 10 – 15 terms.

A series is convergent, if, as the number of terms increases, the sum tends to a definite limit. In certain circumstances, when the relationship between the terms is known, it is possible to make a series converge faster by algebraic manipulation. This is discussed in section 5.03.3 below.

## 5.02 Exponential and related series

The exponential series is very important, not only because of its wide application, but also because other useful series are related to it and it is one of the few series that converge for all values of the variable.

Of course, practically every machine has an adequate built–in exp() program so that the techniques given below will only rarely be used for the calculation of the exponential function itself, but the methods are applicable to several types of power series and, by using exp() for demonstration, it is much easier to illustrate the accuracy achieved.

We have:

$$\exp(x) = 1 + x + x\hat{}2/2 + x\hat{}3/3! + \ldots \qquad 5.01$$

and

$$(1 + x/n)\hat{}n \text{ tends to } \exp(x) \text{ as } n \text{ tends to } \infty \qquad 5.02$$

Relation 5.01 is most useful when $x$ is less than about 0.1, but converges very slowly when $x$ is larger than 1. Table 5.01 gives the precision achieved using relation 5.01 for various values of $x$ with 5 and 10 terms of the series.

By means of a simple programming technique, relation 5.02 can be used for values of $x$ up to 5 with limited accuracy. Table 5.02 shows the results for values of $x$ with $n$ = 1024 = 2^10

The pseudo–code for this method is as follows:

```
EXPOM.PSD(x,k,n)
comment: subroutine to calculate exp(x) using (1+x/n)^n.
Initially, n is chosen to be 2^k (k is an integer), the
variable z is set equal to (1+x/n). Then z is repeatedly
replaced with its square k times. The final value of z is
the required approximation. A major objection to the
method is that it is subject to rounding error because x/n
is much less than 1 and the least significant digits are
lost in the first addition and in the successive
multiplications. The method is, however, quite fast as it
requires only k multiplications, 1 division and 1
addition. Thus, for quick estimates it has some
advantages. endcomment.

   CONST: INT k, n;
   VAR: INT i; REAL x, z;
   BEGIN z = 1+x/n;
      DO (i = 1; i = k; i = i+1;)
          z = z * z;
      CONTINUE
     RETURN (z)
   END
```

**Table 5.01** Calculation of exp(x) using the exponential series

The algorithm POLYNOM.PSD was used for the calculations with a precision of 5 parts in 1E11.

Where no value is given for the error, it is less than the precision.

| *x* | *Calc. Value 5 Terms* | *Error ppm* | *Calc. Value 10 Terms* | *Error ppm* | *10 Digit Approx.* |
|---|---|---|---|---|---|
| 0.1 | 1.105171 | 0.08 | 1.105170918 | – | 1.105170918 |
| 0.2 | 1.22140 | 2.3 | 1.221402758 | – | 1.221402758 |
| 0.3 | 1.34984 | 15.8 | 1.349858808 | – | 1.349858808 |
| 0.4 | 1.49173 | 61.2 | 1.491824698 | – | 1.491824698 |
| 0.5 | 1.64844 | 172.1 | 1.648721270 | 0.0001 | 1.648721271 |
| 0.6 | 1.82140 | 394.5 | 1.822118799 | 0.0008 | 1.822118800 |
| 0.7 | 2.01217 | 785.5 | 2.013752699 | 0.004 | 2.013752707 |
| 0.8 | 2.22240 | 1411 | 2.225540897 | 0.014 | 2.225540928 |
| 0.9 | 2.45384 | 2344 | 2.459603006 | 0.043 | 2.459603111 |
| 1.0 | 2.70833 | 3660 | 2.718281526 | 0.11 | 2.718281828 |
| 2.0 | 7.0 | (5.3%) | 7.388712522 | 46.5 | 7.389056099 |

**Table 5.02** Calculation of (1 + *x*/(2^*k*))^(2^*k*) as an estimate of exp(*x*)

The algorithm EXPOM.PSD was used with *k*=10 (10 Multiplications) and *k*=11 (11 Multiplications)

| *x* | *Calc. Value k = 10* | *Error %* | *Calc. Value k = 11* | *Error %* | *10 Digit Approx.* |
|---|---|---|---|---|---|
| 1 | 2.71696 | 0.05 | 2.7176180 | 0.02 | 2.718281828 |
| 2 | 7.3747 | 0.19 | 7.381848 | 0.1 | 7.389056099 |
| 3 | 19.998 | 0.44 | 20.0415 | 0.22 | 20.08553692 |
| 4 | 54.17 | 0.78 | 54.3856 | 0.39 | 54.59815003 |
| 5 | 146.62 | 1.2 | 147.512 | 0.61 | 148.4131591 |
| 10 | 20983.4 | 4.7 | 21497.0 | 2.4 | 22026.46579 |

Examination of tables 5.01 and 5.02 shows that, for values of *x* less than 1, the relation 5.01 gives very satisfactory results with 10 terms requiring a total of 8 multiplications, 1 division and 9 additions. The relative error, however, increases drastically when $x$ is greater than 1. Relation 5.02 performs better for larger values of x with roughly the same computing overhead. (Addition is generally faster than multiplication.)

This raises an important consideration for functions which are often used and incorporated either into a library file or into a subroutine of a particular program. Neither of the relations 5.01 or 5.02 is satisfactory for values of $x$ greater than 1 if high precision is required, but such values must be available when called by the running program without incurring too great a time penalty. Now that memory is relatively cheap and available, this problem can be overcome by using a look–up table and interpolation. The following example illustrates this method.

## 5.02.1 Interpolation for exp($x$)

Any value of $x$ can be written as $k \pm z$, where k is an integer and $z$ is less than or equal to 0.5. Table 5.01 above shows that for arguments less than 0.5, relation 5.01 converges very quickly and can be used to give accurate results. If all integer values of the arguments up to, say 100, are pre–calculated and stored in an array, then the value for any argument $k+z$ can be calculated from the equation:

$$y = \exp(k + z) = \exp(k) * (1 + z + z\text{^}2/2 + .....) \qquad 5.03$$

It should be noted here that, if POLYNOM is used, the number of terms is fixed and the relative accuracy (but not the precision) will vary slightly with the values of $k$ and $z$. As POLYNOM is fast and, in any case, the final product is probably limited by the machine precision, this is not significant. Larger values of $x$ can be dealt with by breaking up the argument and using repeated multiplication if the arithmetic system can accommodate the larger numbers.

In addition, no test is employed to determine whether sufficient accuracy has been achieved. For well–known functions this is not usually necessary and it is not easy to include a test with POLYNOM because the number of terms is fixed before the calculation. In most cases the speed of POLYNOM allows the number of terms to be increased sufficiently to take care of the required precision for the worst cases, and the slight waste of time in the easier values is offset by the gain in dispensing with the test statements.

## 5.02.2 Functions related to the exponential

The functions sin($x$), cos($x$), sinh($x$) and cosh($x$) are obtainable from the exponential by the following equations: ($j = \sqrt{(-1)}$);

$$\sin(x) = (\exp(j*x) - \exp(-j*x))/2 \qquad 5.04$$

$$\cos(x) = (\exp(j*x) + \exp(-j*x))/2 \quad 5.05$$

$$\sinh(x) = (\exp(x) - \exp(-x))/2 \quad 5.06$$

$$\cosh(x) = (\exp(x) + \exp(-x))/2 \quad 5.07$$

which lead to the expansions:

$$\sin(x) = \Sigma\ (j*x)\verb|^|(2*k+1)/(j*(2*k+1)!);\ k = 0 \text{ to } k = \infty \quad 5.08$$

$$\cos(x) = \Sigma\ (j*x)\verb|^|(2*k)/((2*k)!);\ k = 0 \text{ to } k = \infty \quad 5.09$$

$$\sinh(x) = \Sigma\ x\verb|^|(2*k+1)/(2*k+1)!;\ k = 0 \text{ to } k = \infty \quad 5.10$$

$$\cosh(x) = \Sigma\ x\verb|^|(2*k)/(2*k)!;\ k = 0 \text{ to } k = \infty \quad 5.11$$

All the terms in each of these expansions are real. Sin($x$) and sinh($x$) are odd functions and cos($x$) and cosh($x$) are even. Because of the $j$ factor in 5.08 and 5.09, the terms of the sin() and cos() expansions are alternating in sign whereas those of cosh() and sinh() are all positive. Therefore, as x becomes greater than 1, sinh() and cosh() increase without limit while sin() and cos() are alternating functions. Table 5.03 shows that, for values of the argument up to 0.02, sin($x$) = $x$ = sinh($x$) to less than 1 ppm. With two terms the same precision is obtained up to $x$ = 0.1 and the error when $x$ = 1 is still less than 1%. The replacement of cos($x$) and cosh($x$) by 1 is, on the other hand, only valid to 1 ppm for $x$ less than or equal to 0.0015, or to 0.1% for $x$ up to $x$ = 0.05, and to 1% up to x = 0.1.

With two terms cos() and cosh() may be replaced by the corresponding expansion with values of $x$ up to 0.07 (1 ppm), $x$ up to 0.3 (0.1 %) and $x$ up to 0.7 (1 %). See Table 5.04.

When $x$ is greater than 14, the difference between cosh($x$) and sinh($x$) is less than 1 ppm and both expressions may be replaced by 0.5*(exp($x$)).

**Table 5.03** Comparison between approximations for sin(x) and sinh(x) up to x = 1 rad (57.3°) for one- and two-term expansions.

| *x* | | *sin(x)* | | | *sinh(x)* | | |
|---|---|---|---|---|---|---|---|
| *rad* | ° | *1 term* | *2 terms* | *6 digit value* | *1 term* | *2 terms* | *6 digit value* |
| 0.01 | 0.57 | 0.01 | 0.010000 | 0.010000 | 0.01 | 0.010000 | 0.010000 |
| 0.02 | 1.15 | 0.02 | 0.019999 | 0.019999 | 0.02 | 0.020001 | 0.020001 |
| 0.03 | 1.72 | 0.03 | 0.029996 | 0.029996 | 0.03 | 0.030005 | 0.030005 |
| 0.04 | 2.29 | 0.04 | 0.039989 | 0.039989 | 0.04 | 0.040011 | 0.040011 |
| 0.05 | 2.86 | 0.05 | 0.049979 | 0.049979 | 0.05 | 0.050021 | 0.050021 |
| 0.06 | 3.44 | 0.06 | 0.059964 | 0.059964 | 0.06 | 0.060036 | 0.060036 |
| 0.07 | 4.01 | 0.07 | 0.069943 | 0.069943 | 0.07 | 0.070057 | 0.070057 |
| 0.08 | 4.58 | 0.08 | 0.079915 | 0.079915 | 0.08 | 0.080085 | 0.080085 |
| 0.09 | 5.16 | 0.09 | 0.089879 | 0.089879 | 0.09 | 0.090122 | 0.090122 |
| 0.1 | 5.73 | 0.1 | 0.099833 | 0.099833 | 0.1 | 0.100167 | 0.100167 |
| 0.2 | 11.46 | 0.2 | 0.198667 | 0.198669 | 0.2 | 0.201333 | 0.201336 |
| 0.3 | 17.19 | 0.3 | 0.295500 | 0.295520 | 0.3 | 0.304500 | 0.304520 |
| 0.4 | 22.92 | 0.4 | 0.389333 | 0.389418 | 0.4 | 0.410667 | 0.410752 |
| 0.5 | 28.65 | 0.5 | 0.479167 | 0.479426 | 0.5 | 0.520833 | 0.521095 |
| 0.6 | 34.38 | 0.6 | 0.564000 | 0.564642 | 0.6 | 0.636000 | 0.636654 |
| 0.7 | 40.11 | 0.7 | 0.642833 | 0.644218 | 0.7 | 0.757167 | 0.758584 |
| 0.8 | 45.84 | 0.8 | 0.714667 | 0.717356 | 0.8 | 0.885333 | 0.888106 |
| 0.9 | 51.57 | 0.9 | 0.778500 | 0.783327 | 0.9 | 1.02150 | 1.02652 |
| 1.0 | 57.3 | 1.0 | 0.833333 | 0.841471 | 1.0 | 1.16667 | 1.17520 |

**Table 5.04** Comparison between approximations for cos(x) and cosh(x) up to x = 0.7 rad (40.11°) for one– and two–term expansions.

| *x* | | *cos(x)* | | | *cosh(x)* | | |
|---|---|---|---|---|---|---|---|
| *rad* | ° | *1 term* | *2 terms* | *6 digit value* | *1 term* | *2 terms* | *6 digit value* |
| 0.001 | 0.06 | 1 | 1.000000 | 1.000000 | 1 | 1.00000 | 1.00000 |
| 0.01 | 0.57 | 1 | 0.999950 | 0.999950 | 1 | 1.00005 | 1.00005 |
| 0.02 | 1.15 | 1 | 0.999800 | 0.999800 | 1 | 1.00020 | 1.00020 |
| 0.03 | 1.72 | 1 | 0.999550 | 0.999550 | 1 | 1.00045 | 1.00045 |
| 0.04 | 2.29 | 1 | 0.999200 | 0.999200 | 1 | 1.00080 | 1.00080 |
| 0.05 | 2.86 | 1 | 0.998750 | 0.998750 | 1 | 1.00125 | 1.00125 |
| 0.06 | 3.44 | 1 | 0.998200 | 0.998201 | 1 | 1.00180 | 1.00180 |
| 0.07 | 4.01 | 1 | 0.997550 | 0.997551 | 1 | 1.00245 | 1.00245 |
| 0.08 | 4.58 | 1 | 0.996800 | 0.996802 | 1 | 1.00320 | 1.00320 |
| 0.09 | 5.16 | 1 | 0.995950 | 0.995953 | 1 | 1.00405 | 1.00405 |
| 0.1 | 5.73 | 1 | 0.995000 | 0.995004 | 1 | 1.00500 | 1.00500 |
| 0.2 | 11.46 | 1 | 0.980000 | 0.980067 | 1 | 1.02000 | 1.02007 |
| 0.3 | 17.19 | 1 | 0.955000 | 0.955336 | 1 | 1.04500 | 1.04534 |
| 0.4 | 22.92 | 1 | 0.920000 | 0.921061 | 1 | 1.08000 | 1.08107 |
| 0.5 | 28.65 | 1 | 0.875000 | 0.877583 | 1 | 1.12500 | 1.12763 |
| 0.6 | 34.38 | 1 | 0.820000 | 0.825336 | 1 | 1.18000 | 1.18547 |
| 0.7 | 40.11 | 1 | 0.755000 | 0.764842 | 1 | 1.24500 | 1.25517 |

### 5.02.3 Example: A thin wire stretched between two points

This example demonstrates the importance of checking the allowable level of approximation at different stages in a calculation and shows that a term which may be neglected in one aspect can be very important in another part of the same problem.

The behaviour of wires hung between posts or pylons has a number of applications such as power cables, telephone distribution, ski lifts, cable transporters and etc. Essentially, the problem reduces to a wire supported at its ends, hanging under its own weight, with the possible addition of an extra load, and subject to a temperature variation. If it can be assumed that the lateral stiffness of the wire is negligible, then the form of the curve that the wire takes is that of the common catenary:

$$y = c*\cosh(x/c) \qquad 5.12$$

where: $y$ is the vertical height of a point on the curve above an origin a distance c below the lowest point; $x$ is the horizontal distance from the origin with a maximum value of $r$ (half the distance between the supports). (see Fig. 5.01).

For a very deep catenary $c$ is small compared with the sag of the curve, but when $c$ is much larger than the sag, the catenary becomes flatter. In the present example we are generally concerned with relatively large values of the parameter $c$, so that y/c is between 1 and 2 up to the maximum value for $x$.

Let $T$ be the tension in the wire at the right hand end,
$r$ be half the distance between the supports,
$l$ be half the unstretched length of the wire,
$h$ be half the extension of the wire under load,
$m$ be the mass of unit length of the wire,
$z$ be the height of the supports above the lowest point,
$g$ be the acceleration due to gravity = 9.81 m/s/s,
$Sw$ be the stress in the wire,
$Y$ be Young's modulus for the wire,
$\rho$ be the density of the material of the wire and
$A$ be its cross-sectional area.

Then we have the following equations:

$$T = A*Y*h/l = A*Sw \text{ (elastic stretching)} \qquad 5.13$$

$$T = m*g*(z + c) \text{ (catenary relation)} \qquad 5.14$$

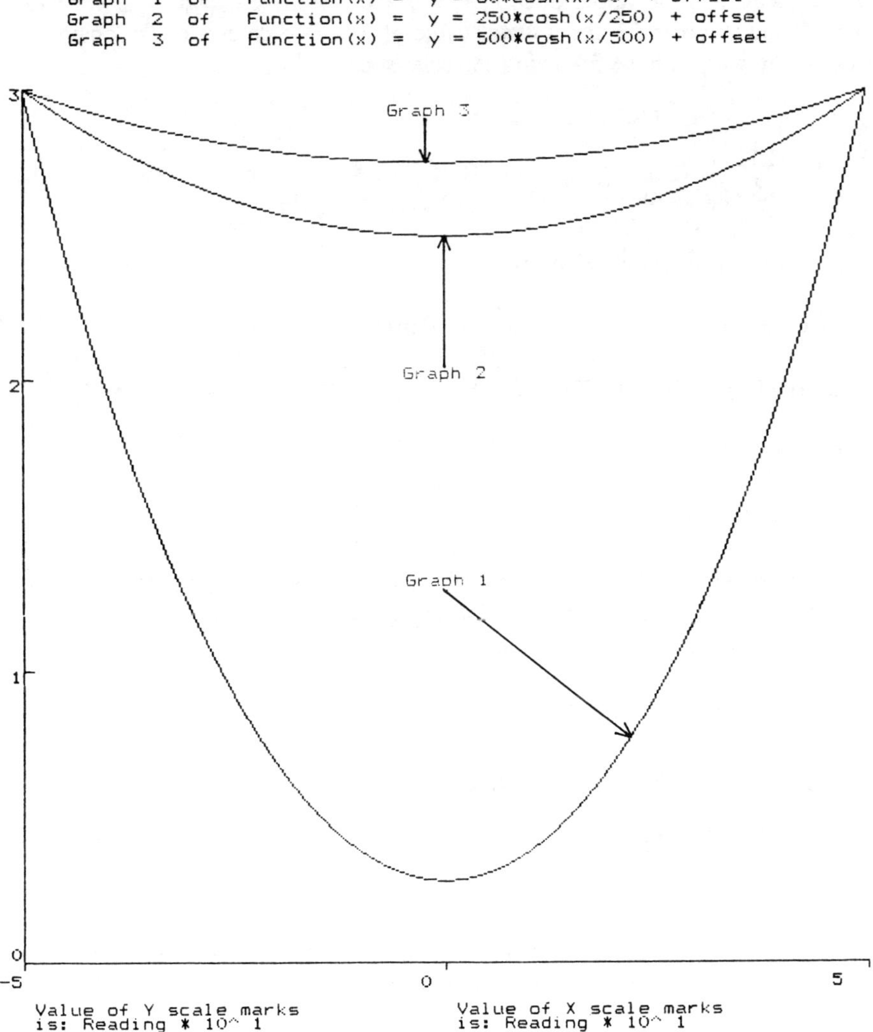


**Figure 5.1** Catenaries with differing values of c

$z + c = y = c*\cosh(r/c)$ (catenary relation) 5.15

$l + h = c*\sinh(r/c)$ (catenary relation, length of arc) 5.16

In these equations, we know that $Sw$ must be, initially, less than the safe working stress for the wire, and the values of $\rho$, $Y$, $A$, $m$, $r$ and $g$ are fixed. Thus, $l$, $c$, $h$, $z$ and $T$ must be found. From 5.13, for a given value of $Sw$, $T$ is determined. Consider a phosphor-bronze wire with the following characteristics:

$A$ = 3.859E–5 $m^2$
$Y$ = 1E11 $N/m^2$
$\rho$ = 8800 kg/cu. m
$m$ = $\rho * A$ = 0.3396 kg/m
$Sw$ = 1E8 $N/m^2$ (about 1/4 of yield stress)

and a distance between supports of 80 m, ($r$ = 40 m)

Therefore $T = A*Sw$ = 3.859E3 N 5.17

From 5.14 and 5.15,

$T/(m*g) = c*\cosh(r/c)$ 5.18

Now, we expect $r/c$ to be less than 1, so that, instead of trying to solve the transcendental equation 5.18, we can use the expansion given in equation 5.11 above. Substituting values and expanding, we get;

$1158.35 = c*(1 + (r/c)^2/2 + ......)$

or,

$1158.35 = c + r^2/(2*c) + .......$ metre. 5.19

First try taking only the first term, then we get;

$c$ = 1158.35 m and $r^2/(2*c)$ = 0.7 m 5.20

The $r^2/(2*c)$ term is about 0.06% of $c$ and, therefore, as a first approximation, the value of $c$ given by 5.20 is satisfactory if the uncertainties in the figures for $T$ and $m$ are considered.

Solving for $z$ in 5.15 and again using the expansion, we get;

$z = r^2/(2*c)$ = 0.7 m 5.21

Note that, in this equation, the c term on each side cancels and we are left with the higher powers. This shows that it is very dangerous to approximate before substituting numerical values unless previous knowledge of the system is available.

The stretched length $l + h$ can now be determined from 5.16 giving;

$l + h = r + r$^3/(6*$c$2) + $r$^5/(5!*$c$^4 ) + ..... 5.22

Here again, to take only the first term is not satisfactory because we would then have the wire length equal to the distance between supports and this is definitely not true.

$r$^3/(6*$c$^2) = .0079 m and $l + h$ = 40.008m 5.23

but, from 5.13 $h/l$ = 1E8/1E11 = .001 5.24

and

$l$ = 39.97 m 5.25

The unstretched length for the chosen stress is slightly less than the distance between the supports, so that the wire must be stressed on installation. It is advisable to ensure that the unstressed length corresponds to the lowest temperature likely to be experienced. Otherwise the contraction occurring when the temperature drops will cause a considerable extra stress in the wire.

Now let us examine the effect of a possible build up of ice on the wire. Suppose that the ice forms a uniform coating 0.5mm thick on the wire. Taking the density of ice as 917 kg/cu. m, the mass per metre of the coating is 0.1729 kg and the new value of $m$ = $m1$ becomes 0.3396 + 0.1729 = 0.5125 kg/m

The equations 5.13–5.16 cannot be so easily solved for the new conditions because the value for $Sw$ can no longer be chosen, but it is set by the external conditions. Expanding equation 5.16 gives

$l + h = r + r$^3/(6*$c$^2) + ..... 5.26

or $1+h/l = r/l$ + $r$^3/(6*$c$^2*$l$) +....... 5.27

Now, $h/l$ is of the order of 0.001 so that $c$ and $l + h$ can, to a first approximation, be taken as constant. Then we have:

$T1 = m1$*$g$*$c$ = 1158.35*0.5125*9.81 = 5824 N 5.28

The total weight of ice on half the wire is:

$W1$ = 0.1729*40*8.81 = 67.84 N 5.29

but the increase in tension is approximately 1965 N

or, increase in tension = $W1*c/r$ 5.30

This property of a flat catenary of increasing the tension in the wire by a significant factor with added load has some advantages and some disadvantages. Where the wire is subject to external variations, the initial form of the curve must be chosen so that the safe working load is never exceeded under the worst conditions likely to be experienced. In general, this means that the catenary must be rather more curved than a simple calculation would indicate.

When equations 5.13 to 5.16 are rewritten neglecting the higher powers of r/c, we get the following set:

Let ang be the angle that the wire makes with the horizontal at the supports, then;

$z = r\text{^}2/(2*c)$ 5.31

$dz/dx = \tan(ang) = r/c = 2*z/r$, (at $x = r$) 5.32

$T = A*Sw$ 5.33

$c = T/(m*g)$ 5.34

$h/l = Sw/Y$ 5.35

$l + h = r + r\text{^}3/(6*c\text{^}2)$ 5.36

(This term cannot be neglected)

From 5.32 it can be seen that the angle ($ang$) is almost constant because, as shown above, $c$ varies very little with change of load although it will change significantly if $l$ changes. (For example if the temperature varies.)

Because the curve is so flat under these conditions, it is very tempting to assume that, when the wire is loaded with a force $F$ at its mid point, the curve can be taken as two straight lines as shown in fig. 5.02. This is never the case, because, from physical considerations, the wire can only be straight if it is inextensible and the distance $2*r$ between the supports cannot then be fixed. Also, in practical situations the slope at the lowest point of the wire must be zero, whereas for the two straight lines, the slope is discontinuous at the lowest point.

If the supports are flexible, however, and the extension of the wire can be neglected compared with the change in r, Fig. 5.02 may be used for the estimation of the tension at the supports, but the above equations are no longer applicable.

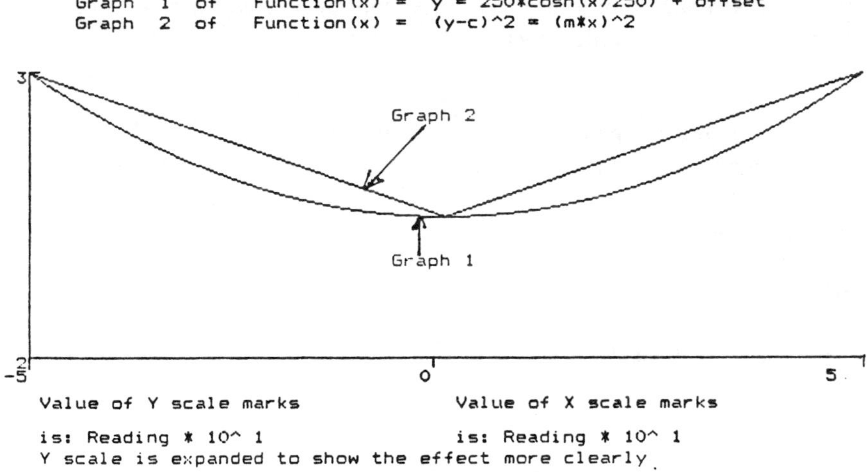


**Fig. 5.02** Catenary and two straight lines

## 5.02.4 Example: Population Growth

In Chapter 3, Figure 3.01, the population growth curve given by:

population at time $t = P = P0/(1 + A*\exp(-m*t))$ 5.37

where $P0$ is the final population and $A$ and $m$ are constants, was used as an illustration of the effects of log-log plots.

Here we shall use it to demonstrate how differing approximations may be used to represent different aspects of a relationship as the independent variable changes.

For most purposes, the expression in 5.37 may be calculated directly, but loss of significance occurs when $A*\exp(-m*t)$ is less than 1.01 or when $A*\exp(-m*t)$ is greater than 100. In addition, near the point of inflection, the curve can be considerably simplified.

When computing systems are available, it is often easier to use the given formulas directly and let the machine do the work. This course will be satisfactory for many purposes, especially if a numerical print out is accompanied by a graphical display. Nevertheless, it is very helpful to be able to use properly derived approximations for particular parts of a curve in order to gain a "feel" for the manner in which a system is changing.

Equation 5.37 is a good example of this type of function as its initial behaviour is not immediately obvious from the form of the expression.

It is convenient to write 5.37 as:

$$(P/P0)\% = F = 100/(1 + A*\exp(-m*t)) \qquad 5.38$$

Put $A = 199$ and m = 0.01 and we have:

$$F = 100/(1 + 199*\exp(-0.01*t))\ \% \qquad 5.39$$

At $t = 0$, the initial population is 0.5% of the final value.

Table 5.05 gives values of $F$ for $t$ in steps of 50 units with the values of the approximation functions for comparison.

When $t$ is between 0 and 250, equation 5.38 may be rewritten as:

$$F = 100*\exp(m*t)/A*(1 + \exp(m*t)/A) \approx 0.5*\exp(0.01*t) \qquad 5.40$$

When $t$ is greater than 750 the approximation is derived as follows:

$F = 100/(1 + f(t))$, where $f(t)$ is less than 1 and we may therefore use the binomial expansion which gives:

$$F \approx 100*(1 - 199*\exp(-0.01*t)) \qquad 5.41$$

The curve for $F$ has a point of inflection at $t = ti$ given by:

$$A*\exp(-m*ti) = 1, \qquad Fi = 50\% \qquad 5.42$$

At a point of inflection, the slope of a curve has a maximum or minimum and, therefore , near this point the equation of the curve itself must approximate to a straight line.

Solving 5.42 for $ti$ gives $ti = 529.33$ 5.43

Near this value then, $F \approx 0.25*(t - 529.33) + 50$ 5.44

Between $t = 450$ and $t = 600$, this relation holds to better than 1%, but the original function is preferred for the parts between $t = 250$ and $t = 450$, and $t = 600$ and $t = 750$.

When such a curve is fitted to measured data, it should be noted that, in the initial stages, it is virtually impossible to distinguish the expression given by 5.37 from a normal exponential growth relationship. Thus, without more knowledge of the characteristics of the system being investigated, extrapolation beyond the limits of the available data is extremely suspect.

This is a general rule, but prediction is, and has been since time immemorial, a favourite sport of the human race. Agreed, it is necessary for planning and estimation purposes, but, unless due allowance is made for adequate error margins, the outcome can easily be disastrous.

**Table 5.05** Values of $F = 100/(1+199*\exp(-0.01*t))$ % for increments of 50 units in $t$ with the relevant approximations:
$F1 = 0.5*\exp(0.01*t)$; $F2 = 0.25\ (t - 529.33)+50$; $F3 = 100*(1 - 199*\exp(-m*t))$

| *t units* | *F%* | *F1%* | *F2%* | *F3%* | *Error%* |
|---|---|---|---|---|---|
| 0 | 0.5 | 0.5 | | | 0. |
| 50 | 0.823 | 0.824 | | | -0.1 |
| 100 | 1.348 | 1.359 | | | -0.8 |
| 150 | 2.203 | 2.241 | | | -1.7 |
| 200 | 3.580 | 3.695 | | | -3.2 |
| 250 | 5.769 | 6.091 | | | -5.6 |
| 300 | 9.168 | 10.043 | | | -9.5 |
| 350 | 14.267 | | | | |
| 400 | 21.529 | | 17.668 | | 17.9 |
| 450 | 31.146 | | 30.168 | | 3.1 |
| 500 | 42.719 | | 42.668 | | 0.12 |
| 550 | 55.149 | | 55.168 | | -0.03 |
| 600 | 66.967 | | 67.668 | | -1.0 |
| 650 | 76.971 | | 80.167 | | -4.1 |
| 700 | 84.641 | | | 81.854 | 3.3 |
| 750 | 90.085 | | | 88.994 | 1.2 |
| 800 | 93.742 | | | 93.324 | 0.43 |
| 850 | 96.109 | | | 95.951 | 0.16 |
| 900 | 97.603 | | | 97.444 | 0.16 |
| 950 | 98.523 | | | 98.510 | 0.01 |
| 1000 | 99.105 | | | 99.097 | 0.008 |

```
Graph 1 of  Function (t) = 100/(1 + 199*exp(-0.01*t))
Graph 2 of  Function (t) = 0.5*exp(0.01*t))
Graph 3 of  Function (t) = 0.25*(t - 529.33) + 50
Graph 4 of  Function (t) = 100*(1 - 199*exp(-0.01*t))
```

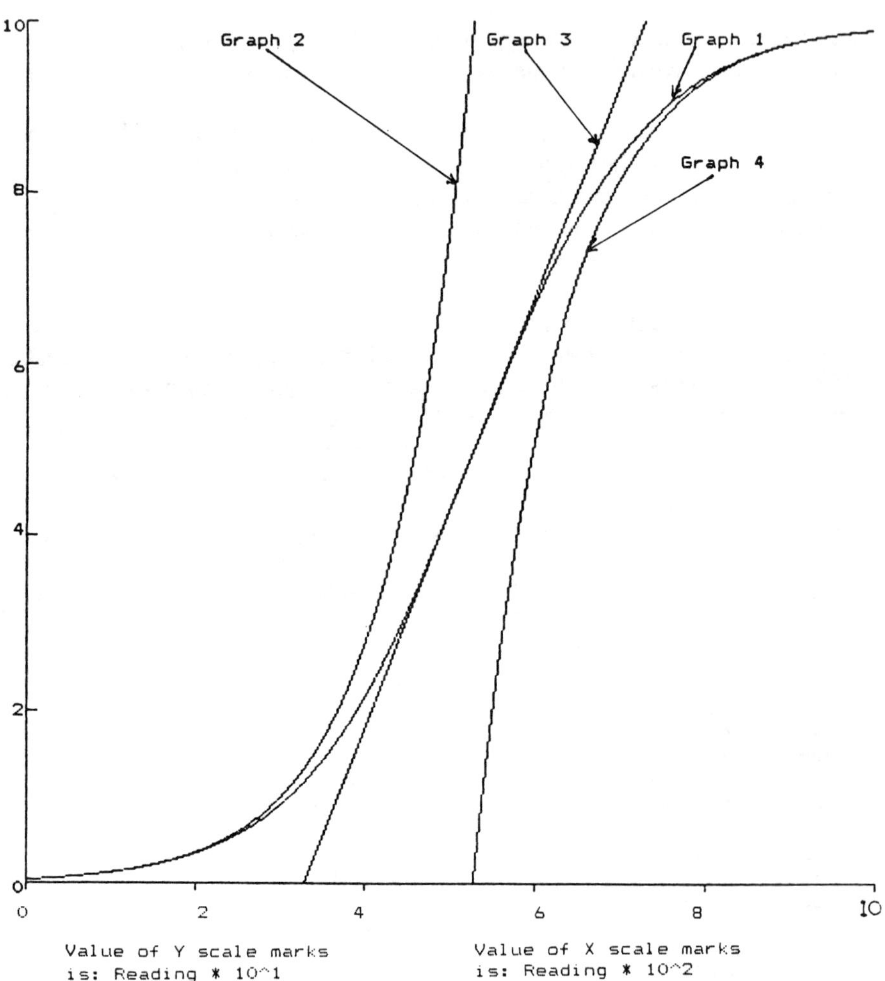


**Figure 5.03** Population Growth Curve with Approximations

Figure 5.03 shows these four relations plotted together on the same axes for comparison purposes.

# 5.03 Curve Representations and Curve Fitting

In Fig. 5.04, the first differential (slope) of each of the equations of Fig. 5.03 is plotted and from these curves it can be seen that, near the changeover points, the relative change in slope is much greater than the relative change in value. This a general rule which holds unless there is a special relationship between the successive component curves representing an original equation. At the corner points where the replacement slopes meet, the slope is discontinuous. As shown in section 5.03.6, it is sometimes possible to find expressions in which the slope as well as the value is similar to that of the target curve.

The replacement expressions for equation 5.38 were obtained by manipulation of the mathematical formula and applying the rules given in section 4.1.1, but the approximations are only valid for specific parts of the relationship. In the particular case of the population curve, the basic equation is not complex enough to warrant further simplification, but, in other cases, it may be necessary to find calculable substitutions for the curve over its whole range.

There are a number of different types of method for this procedure and a list of the more important ones follows.

Taylor's series (expansion about a fixed value)

General power series

Polynomial curve fitting (partial or total)

Fourier series (mostly for periodic functions)

Replacement by a more familiar curve (partial or total)

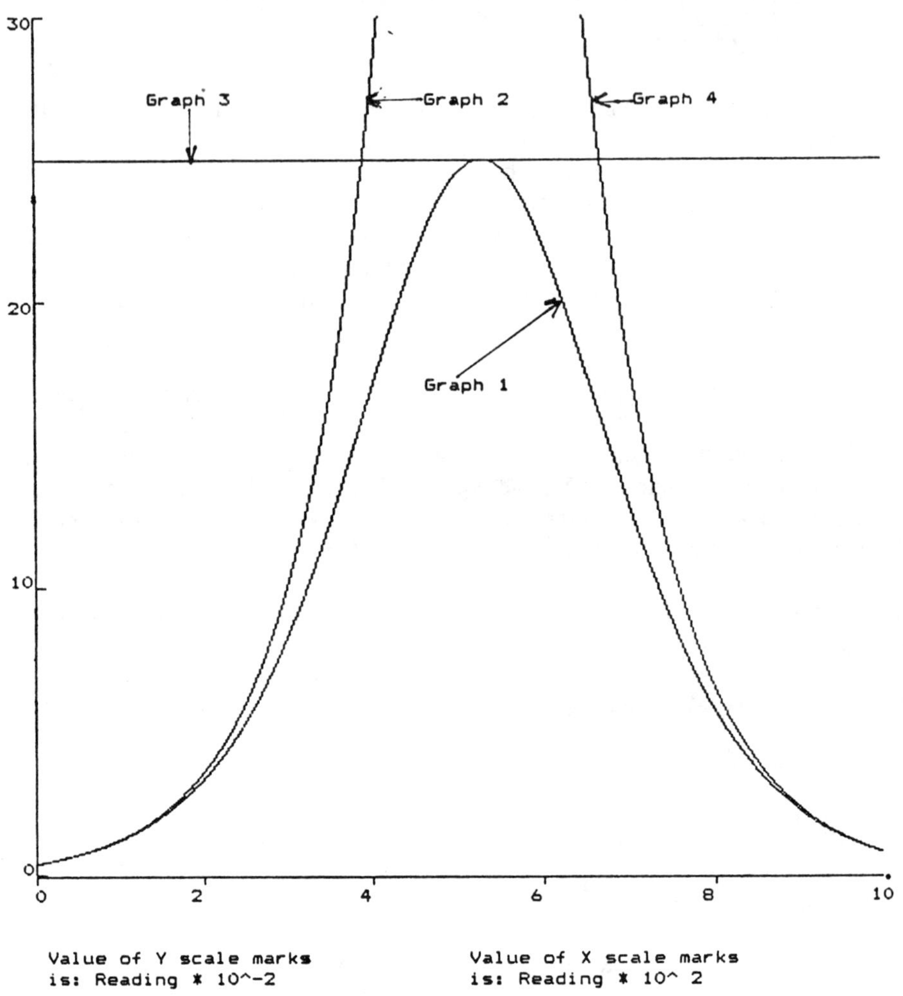


**Figure 5.04** Population Growth: Slopes

## 5.03.1 Taylor's Series

The usual form for Taylor's Series is:

$$f(x0+h) = f(x0) + \Sigma(h\char`^n*(f\char`^n(x0))/n!) + Rn \qquad 5.45$$

where $f\char`^n(x0)$ is the $n$th derivative of $f(x)$ at $x0$ and $Rn$ is the value of the remainder after $n + 1$ terms.

When x is not zero and the absolute value of h is less than x, it is good practice, in accordance with rule 5 of section 4.1 1 to write this relation as:

$$f(x0(1 + h/x0)) = f(x0) + \Sigma(h\char`^n*(f\char`^n(x0))/n!) + Rn \qquad 5.46$$

whence the relative extent of $h$ can easily be seen.

One of the great advantages of Taylor's Series is that, if the form of the curve is not known, it is still possible to use the expression if the relevant derivatives can be estimated.

In general engineering practice the use of the remainder is not common, but, with modern techniques, it takes very little extra trouble to include it and it gives a useful check on validity.

Let $Rna$ and $ha$ be the absolute values of $Rn$ and $h$ respectively, and $Max$ be the maximum of the absolute value of $f(x)$ between $x - h$ and $x + 2*h$, then we have, approximately:

$$Rna < ha\char`^(n+1)*Max/(n+1)! \qquad 5.47$$

or, more approximately:

$$\text{Era} \approx ha\char`^(n+1)/(n+1)! \qquad 5.48$$

The following simple example illustrates how this works.

Suppose we have two values from a tabular function

| $x$ | $y$ |
|---|---|
| 0.9 | 0.783 |
| 1.0 | 0.841 |

and we wish to find the value at $x = 0.96$ using a linear model.

Then from 5.45: f(0.96) $= 0.783 + 0.06*(0.841 - 0.783)/0.1$
$= 0.8178$ 5.49

and the estimated percentage error
$= Era*100$
$= 0.1*0.1*100/2$
$= 0.5\%$. 5.50

These values were taken from a table of sines and the actual value at $x = 0.96$ rad is 0.8192 giving an error of 0.17% which is well within the estimated value. This example demonstrates that, at the given value and with the given interval, a sine curve may be replaced by a straight line to better than 0.5%.

Now the estimation of higher differentials from tabulated data becomes more unreliable as the order of the differential increases as discussed in section 6.06, so that the estimation of error of interpolated values from tabulated data using the Taylor remainder is not recommended for models of higher order than two. If, however, the form of the curve is known or can reasonably be inferred, then it may be possible to use higher orders. Unless the data are widely spaced, it is seldom necessary to use a model of higher order than two, requiring only the estimation of a second differential. If the data are widely spaced, then a curve should be fitted to them before attempting to interpolate.

## 5.03.2 General Power Series

The Binomial expansion, the exponential–type series and Taylor's Series are all examples of power series, but they are of little use when the variable is such that the series converges slowly or not at all. Functions which are the quotients or products of two other series can sometimes also be expressed as a series over a limited range. The general term of such an expression may be very complex and, for practical purposes, it is usually preferable to calculate the first few coefficients using the polynomial multiplication and division techniques given in section 4.8. However, because the series do not end, the method must be modified to start from the lowest power in the case of division.

```
SUBROUTINE DIVSER.PSD(n,k,coef1(n),coef2(n),coef3(n),rm
(n+1))

comment: both series should start with a constant term
(zero if necessary) and end with the same power of x. The
quotient coefficients are returned in coef3() and the
remainder in rm(). The n remainder coefficients should
normally be zero, but there may be some residuals due to
round-off error. The form of the coefficients produced may
be unfamiliar at first sight as text-books generally give
```

them as fractions, but, as the method is intended to give coefficients for further calculation, there seemed little point in adding the extra complication. If the denominator series starts with a power of x greater than zero, k, say, the algorithm moves the divisor to that power and each term of the resulting series must be divided by x^k, but it is then very important to realise that the series is then quite invalid for x = 0 and not very useful for 0 < x < 0.01 unless the coefficients of the negative powers of x are small enough. endcomment.

```
    CONST: INT n; REAL coef1(n),coef2(n);
    VAR: INT i,j,k; REAL u,coef3(n),rm(n);
   BEGIN
      DO (i = 0; i = n; i = i+1;)
          rm(i) = coef1(i); coef3(i) = 0;
      CONTINUE
          k = 0; u = coef2(0);
            WHILE (coef2(k) = 0)
                   u = coef2(k+1); k = k + 1;
            ENDWHILE
          DO (i = 0; i = n1; i = i+1;)
             coef3(i) = rm(i)/u;
               DO (j = i; j = n1; j = j+1;)
                  rm(j) = rm(j) - coef3(i)*coef2(j-i);
               CONTINUE
          CONTINUE
       RETURN (k, coef3(),rm())
   END
```

**Example 5.03.2.1**

Using the above algorithm we can calculate the first nine values of the expansion of tan(x) given the expansions for sin(x) and cos(x).

$$\sin(x) = 0+x+0-x^3/3!+0+x^5/5!+0-\ldots$$
$$\cos(x) = 1+0-x^2/2!+0+x^4/4!+0-x^6/6!+0\ldots$$

When these values up to the ninth power are substituted into the algorithm we get:

$$\tan(x) = x + 0.333333*x^3 + 0.133333*x^5 + 0.053968*x^7 + 0.021869*x^9+.. \qquad 5.51$$

As this series converges so slowly, it is only useful when the value of $x$ is such that the first two terms give a satisfactory result, but it is a good example for demonstration because the coefficients can be checked against the known values.

## 5.03.3 Accelerated convergence of power series

There are a number of tricks that can be used to accelerate the convergence of a series providing the terms are decreasing. In equation 5.51, for example, if $x$ is greater than 1, the terms increase, initially, before eventually decreasing. In such cases the series should be summed until the terms stop increasing before applying the accelerating methods.

### 5.03.3.1 Aitken's delta squared process

This technique can be applied when the successive terms are decreasing approximately geometrically.

Let $S(k)$ be the partial sum at the kth power
$S(k+j)$ be the partial sum at the (k + j)th power
*Sum* be an improved value

Then we have:

$$Sum = S(k+j) - (S(k+j) - S(k))\text{^}2/(S(k+j) - 2*S(k) + S(k-j)) \qquad 5.52$$

Applying this method to 5.51 with x = 1, we get:

$$\begin{aligned} Sum &= S(9) - (S(9) - S(7))\text{^}2/(S(9) - 2*S(7) + S(5)) \\ &= 1.542503 - (1.542503 - 1.520634)\text{^}2/(-0.0032099) \\ &= 1.557402 \qquad 5.53 \end{aligned}$$

The value of tan(1) to 6 decimal places is 1.557407, so that the error in this case is about 3 ppm. To achieve this accuracy with the original series would take about 25 odd power terms. For $x$ larger than 1, the advantage is even greater if the first partial sum is taken when the terms begin to decrease.

### 5.03.3.2 Remainder estimation

If it is possible to estimate a remainder after $k$ terms, it is then necessary only to add this remainder to $S(k)$ in order to obtain an improved result. The following method is based on lectures given to engineering students by a friend and former colleague Harry Fairbrother (Bib.5) and it is reproduced with permission. It is related to the technique of superlinear convergence discussed by M. A. Wolfe in Bib.4.

Let the terms of the series be represented by $a(i)$ where $i$ starts at 0 and the remainder after k terms be $R(k)$. For k = 5 we have:

$$S(k) = a(0) + a(1) + a(2) + a(3) + a(4) + a(5) \quad 5.54$$

and

$$Sum(k) = S(k) + R(k) \quad 5.55$$

Now, if the series is to converge, then $a(k + 1)/a(k)$ must be less than or equal to a limit $lm$ say, as $k$ tends to infinity. Assuming that $a(k + 1)/a(k)$ tends to $lm$, we get:

$$R(k) = a(k)/(1 - lm) \quad 5.56$$

and the $R(k)$ tend to the true remainder as $k$ becomes larger. A new sequence can be set up as follows:

$$\begin{aligned} Sum(i) &= S(i - 1) + a(i)/(1 - a(i)/a(i - 1)) \\ &= S(i - 1) + a(i)*a(i - 1)/(a(i-1) - a(i)) \end{aligned} \quad 5.57$$

The terms can then be regrouped as:

$$\begin{aligned} &Sum(1),(Sum(2) - Sum(1)),(Sum(3) - Sum(2)), \\ &(Sum(4) - Sum(3)),\text{etc} \end{aligned} \quad 5.58$$

leading to a series beginning with $Sum(1)$ and continuing with the successive differences. The final term of this series is then $Sum(k) - Sum(k - 1)$; and we therefore have:

$$Sum = Sum(k - 1) + (Sum(k) - Sum(k - 1))*a(k - 1)/(a(k - 1) - a(k)) \quad 5.59$$

This is one of the fortunate cases where the practice is a lot simpler than the derivation and, in the algorithm given below, the only data needed are $k$ and the $a(i)$

```
SUBROUTINE ACCSER.PSD(k,a(k),sum)

comment: this routine assumes that the series to be evaluated
has decreasing terms in which the successive ratios a(i)/a
(i-1) approach a limit. The terms must be uniformly
decreasing and a series such as sin(x) with alternate zeros
should be closed up to comply with this requirement. A new
series is formed from the original by adding a remainder to
the partial sums and then taking successive differences. The
integer k is a term counter and is not directly related to the
power of x. The terms run from 0 to k. The final estimate is
returned as sum, which is formed from the penultimate
modified sum and a remainder. If required, the normal sum is
available as s for comparison. endcomment.

   CONST: INT k; REAL a(k);
```

```
VAR : INT i; REAL r,s,sm(k),sum
 BEGIN
    s = a(0);
       DO (i = 1; i = k; i = i+1;)
          r = a(i-1)/(a(i-1)-a(i));
          sm(i) = s+r*a(i); s = s+a(i);
       CONTINUE
          sum = sm(k-1)+(sm(k)-sm(k-1))*r;
    RETURN (sum)
 END
```

**Example 5.03.3.3**

Consider the series:

$$\log(1+x) = x - x^2/2 + x^3/3 - x^4/4 + x^5/5 - x^6/6 \ldots. \qquad 5.60$$

For $x = 0.8$,
the 6 digit value is 0.587787
the value from 5.60 is 0.570112
and from the algorithm it is 0.587844 with an error of 57 ppm

To obtain this precision from 5.60 would require about 30 terms.

# 5.03.4 Polynomial curve fitting

There are two aspects to fitting a polynomial to a set of points: (a) the points may be derived from a known expression or, (b) they may result from measured data.

In the first case, the fit can be exact at certain points (within the round off error), but in the second case the fitted curve represents a regression polynomial corresponding to a least squares fit.

There are a number of differing methods available for curve fitting and they are described by Press et al (Bib 1), Lawson (Bib 6), Luke (Bib 7) and many others. Only one system, using orthogonal polynomials, has been selected for presentation here because it is general enough to cover a large proportion of the possible applications and it lends itself to relatively easy attachment to more comprehensive programs. A major advantage of the method is that the succesive orders of the polynomial fit are independent and if, say, a fifth order polynomial has been computed, but a further refinement is required, then the sixth one can be added without altering the previous five.

The technique is well described in Chatfield (Bib 8) and a program for $n+1 = 19$ points, and up to a ninth order curve, is given as the pseudo–code algorithm POLYFIT.PSD which illustrates the method and is powerful enough for most purposes. For different sample sizes, and up to sixth order polynomials, the program must be modified to substitute the appropriate functions and values as tabulated in Pearson and Hartley (Bib 9) or Fisher and Yates (Bib 10) where the maximum sample size is 52 points. It should be noted that the coefficients are strictly confined to the number of points and the highest order selected. They must not be used for a different set. Thus care must be taken when the values are entered in the program. The data (individual $y$ values corresponding to each $x$) must be equally spaced on the $x$ axis and it is recommended that an odd number (n+1) of x points be chosen. This makes the determination of certain parameters easier when a curve is fitted to an histogram derived from statistical data. This aspect is further discussed in Chapter 7.

It may be objected that it is not justified to fit a ninth order polynomial to as few as 19 points, but, when the equation to be replaced has alternate zero terms, the ninth order consists effectively of only 5 separate powers if the fit is symmetrical, but of 9 coefficients when it is not. Table 5.09 shows the difference between two separate approximations to the same function.

Certainly, if the data are subject to severe statistical fluctuations, then it is possible to fit a completely false higher order curve to these data, but the methods of Chapter 7 can help to resolve some of the problems associated with such cases.

The function chosen for Table 5.09 is the density function of the Normal probability distribution: ( $y = \exp(-x*x/2)/(\sqrt{(2*\pi)})$). If the expression is fitted from –3.357 to 3.375 (19 values spaced at 0.357 intervals) the error is less than 0.001 over the range. When the interval is reduced to 0.25 and the $x$ range is taken from 0 to 4.5 the fit is better than 1E–4 over the range.

**Warning**

The polynomial fit is valid only within the range over which it is taken. Any attempt to extrapolate outside the range is extremely hazardous and totally unjustified, especially if the parent curve is not itself a polynomial. This characteristic is demonstrated by Table 5.09. This warning applies especially to attempts to fit curves to financial data on a time axis. The author and publisher can accept absolutely no responsibility for losses occasioned by predictions made on the bases of these programs.

The validity of the fit for data subject to statistical variation may be determined by examining the approriate variance ratios using the Fisher's $F$ distribution. If two successive values of this parameter, calculated from the residual sums of squares provided by the program, are less than the tabulated values for a selected significance and the relevant number of degrees of freedom, the fit can be stopped at the last level. In practice, because of the method for finding the coefficients, it is easier to determine the complete set of coefficients and neglect those which are shown to be redundant.

Strictly speaking, if the data are derived from an equation and not from statistical data, the use of regression is not really valid, but, in practice it seems to work quite well if the parent equation is reasonably well–behaved. If, however, the data are not derived from a polynomial or rational function, Fisher's $F$ test does not work and it is better to use the residuals directly.

Table 5.06 gives the orthogonal polynomials for a sample size of 19 and Table 5.07 shows the corresponding values from which the initial coefficients are calculated. The procedure is explained by the comments in the algorithm. If a different system is required, Tables 5.06 and 5.07 must be reconstructed from the data given in the references mentioned above. It is important to note that in these references the functions have a multiplying factor which makes it possible to have integer values in the tables. Tables 5.06 and 5.07, however, already incorporate this factor and this should be taken into account when making any modifications.

If it is impossible to arrange for the data to be evenly spaced, the following trick may sometimes be used if the $y$ values are varying smoothly and not too violently. It is quicker and much less prone to mistakes than trying to fit the points directly with an ordinary regression program.

Suitable $x$ values ($X$) are chosen to span the given range and, using the routine, a parabola is fitted to the first three available points (*x0,x1,x2*) and a new $y$ calculated at the first $X$. This is then repeated for the next three (*x1,x2,x3*)and so on until the end of the range. The curve obtained in this way will, of course, not be so accurate, but it is often good enough.

**5.03.4.1** `SUBROUTINE PARAFIT.PSD(x(2),y(2),a(2))`

```
comment: a routine to fit a parabola to three points x(i),
y(i) (i = 0 to i = 2). endcomment.

Let the equation of the parabola be:

    y = a0 + a1*x + a2*x^2                                5.61

The three values a(0), a(1) and a(2) are returned.

   CONST: INT i; REAL x(2),y(2);
   VAR  :  INT j; REAL a(2),z(3),w(1);
```

```
BEGIN
        i = 3;
     z(0)  = x(0)-x(1); z(1)  = x(1)-x(2);
     z(2)  = x(0)-x(2); z(3)  = x(0)+x(1);
     w(0)  = y(0)-y(1); w(1)  = y(1)-y(2);
     a(2)  =  (w(0)/z(0)-w(1)/z(1))/z(2);
     a(1)  =  (w(0)/z(0))-(a(2)*z(3));
     a(0)  = y(0)-a(1)*x(0)-a(2)*x(0)*x(0);
    RETURN(a())
 END
```

**Table 5.06** Orthogonal Polynomials for a sample size of n+1 = 19 and order k = 9.

cf(i,j): coefficients of the f(i,z) for the tabulated powers of the dummy variable z = (x–xm)/d

| *i* \ *j* | 0 | 1 | 2 | 3 | 4 | 5 | 6 | 7 | 8 | 9 |
|---|---|---|---|---|---|---|---|---|---|---|
| 0 | 1 | | | | | | | | | |
| 1 | | 1 | | | | | | | | |
| 2 | –30 | | 1 | | | | | | | |
| 3 | | –269/6 | | 5/6 | | | | | | |
| 4 | 396 | | –535/12 | | 7/12 | | | | | |
| 5 | | 1393/30 | | –59/24 | | 1/40 | | | | |
| 6 | –1320 | | 4783/15 | | –263/40 | | 11/120 | | | |
| 7 | | –9605/28 | | 8627/240 | | –11/12 | | 11/1680 | | |
| 8 | 35 | | –150713/10080 | | 373/384 | | –133/6720 | | 5/8! | |
| 9 | | 7429/126 | | –189445/18144 | | 25359/51840 | | –101/12096 | | 17/9! |

In each line *f*(*i*) is to be interpreted as:

*f*(*i*,*z*) = *cf*(*i*,0) + *cf*(*i*,1)**z* + *cf*(*i*,2)**z*^2 + ....... 5.62

To avoid overpopulating the table zero entries are represented by blank spaces.

**Table 5.07** Orthogonal Polynomials for a sample size of $n+1$ = 19 and order $k$ = 9. $zf(i,z)$: values of the $f(i,z)$ for each of the 19 points of the dummy variable $z = (x-xm)/d$

| z | *0* | *1* | *2* | *3* | *4* | *5* | *6* | *7* | *8* | *9* |
|---|---|---|---|---|---|---|---|---|---|---|
| –9 | 1 | –9 | 51 | –204 | 612 | –102 | 1326 | –306 | 17 | –17 |
| –8 | 1 | –8 | 34 | –68 | –68 | 68 | –1768 | 646 | –51 | 68 |
| –7 | 1 | –7 | 19 | 28 | –388 | 98 | –1222 | 86 | 21 | –67 |
| –6 | 1 | –6 | 6 | 89 | –453 | 58 | 234 | –411 | 40.5 | –37 |
| –5 | 1 | –5 | –5 | 120 | –354 | –3 | 1235 | –425 | 7.5 | 42.5 |
| –4 | 1 | –4 | –14 | 126 | –168 | –54 | 1352 | –97 | –28.5 | 56 |
| –3 | 1 | –3 | –21 | 112 | 42 | –79 | 729 | 267 | –34.5 | 3.5 |
| –2 | 1 | –2 | –26 | 83 | 227 | –74 | –214 | 427 | –10.5 | –49 |
| –1 | 1 | –1 | –29 | 44 | 352 | –44 | –1012 | 308 | 21 | –49 |
| 0 | 1 | 0 | –30 | 0 | 396 | 0 | –1320 | 0 | 35 | 0 |
| 1 | 1 | 1 | –29 | –44 | 352 | 44 | –1012 | –308 | 21 | 49 |
| 2 | 1 | 2 | –26 | –83 | 227 | 74 | –214 | –427 | –10.5 | 49 |
| 3 | 1 | 3 | –21 | –112 | 42 | 79 | 729 | –267 | –34.5 | –3.5 |
| 4 | 1 | 4 | –14 | –126 | –168 | 54 | 1352 | 97 | –28.5 | –56 |
| 5 | 1 | 5 | –5 | –120 | –354 | 3 | 1235 | 425 | 7.5 | –42.5 |
| 6 | 1 | 6 | 6 | –89 | –453 | –58 | 234 | 411 | 40.5 | 37 |
| 7 | 1 | 7 | 19 | –28 | –388 | –98 | –1222 | –86 | 21 | 67 |
| 8 | 1 | 8 | 34 | 68 | –68 | –68 | –1768 | –646 | –51 | –68 |
| 9 | 1 | 9 | 51 | 204 | 612 | 102 | 1326 | 306 | 17 | 17 |

The corresponding sums of squares sq($i$) are:

| 19 | 570 | 13566 | 213180 | 2288132 | 89148 | 24515700 | 245170 | 16387.5 | 41055 |
|---|---|---|---|---|---|---|---|---|---|

**Table 5.08** Values of the $F$ distribution for a significance level of 0.05 and for 1 and 9 to 17 degrees of freedom.

| | *f05(1,m)* | | | | | | | | |
|---|---|---|---|---|---|---|---|---|---|
| *m* | 9 | 10 | 11 | 12 | 13 | 14 | 15 | 16 | 17 |
| *f05(1,m)* | 5.12 | 4.96 | 4.84 | 4.75 | 4.67 | 4.60 | 4.54 | 4.49 | 4.45 |

**5.03.4.2** SUBROUTINE POLYFIT.PSD (n,k,x(n),y(n),a(k),b(k), c(k),r(k),ff(k))

comment: the 19 y values corresponding to the equally spaced x points are assumed to be stored in an array y(l) (l = 0 to 18) and the x data in array x(l). The 9*19 values of f(i,z(l)) given in table 5.07 are contained in array zf(i,l) and the coefficients of the nine functions f(i,z) in array cf(i,j); (i,j = 0 to 9): d and xm are the spacing, and the mean, of the x points respectively.

Then we have:

$$z = (x-xm)/d$$

and $y = a(0) + a(1)*f(1,z) + a(2)*f(2,z) + \ldots\ldots\ldots$ 5.63

Equation 5.63 is effectively a polynomial in *z*, but the coefficients of the individual powers of *z* are distributed throughout the *f(i,z)* and must be subsequently disentangled.

The program first determines *d* and *xm* and then calculates the *a*() coefficients from the following relation.

$$a(i) = \sum y(l)*zf(i,l)/sq(i); \quad l= (0 \text{ to } n); \quad i = (0 \text{ to i}) \qquad 5.64$$

Next the *b*() coefficients for the polynomial in *z* are found, and finally, the *c*() values for the replacement polynomial in *x*.

At each level the total residual sum of squares is found from the relation

$$r(j) = \sum (y-ym)^2 - \sum a(i)*a(i)*sq(i); \ i = 1 \text{ to j} \qquad 5.65$$

and the contribution from each level from

$$ff(j) = a(j)*a(j)*sq(j) \qquad 5.66$$

Fishers *F* value is then given by

$$F = (n-j)*ff(j)/r(j) \qquad 5.67$$

where *n-j* is the number of degrees of freedom of *r(j)*.

For orders greater then 5, *r(j)* is usually too small to be reliably used in 5.67 and it is better to estimate the residual as a percentage of (*y-ym*)^2. If required, the calculation of the *F* values can easily be incorporated in the program, but it is not recommended owing to the rapid build-up of rounding error as the order increases. endcomment

```
CONST: INT k,n;  REAL cf(k,k),zf(k,n),sq(k),d,ysq;
VAR:   INT i,j,l; REAL a(k),b(k),c(k),xm,ym,yym,;
       INT m;   REAL bin(k,k),ff(k),r(k)
                     temd,temx,ff1;
BEGIN
  ym = 0; yym = 0; b(0) = 0; c(0) = 0;
  IF (n MOD 2 = 0)
     xm = x(n/2);
  ELSE
     xm = (x(0) + x(n))/2;
  ENDIF
    d = (x(n) - x(0))/n;
   DO (l = 0; l = n; l = l + 1;)
     ym = ym+y(l);
     yym = yym+y(l)*y(l);
   CONTINUE
     a(0) = ym/sq(0); ff(0) = 0;
     ysq = yym-ym*ym/(n+1); r(0) = ysq;
      DO (i = 1; i = k; i = i+1;)
        a(i) = 0; b(i) = 0; c(i) = 0;
           DO (l = 0; l = n; l = l+1;)
              a(i) = a(i)+y(l)*zf(i,l);
            CONTINUE comment: a coefficients;endcomment
       ff1 = a(i);
        a(i) = a(i)/sq(i);
        ff(i) = ff1 * a(i);
        r(i) = r(i-1)-ff(i);
      CONTINUE
       temd = 1;
      DO (j = 0; j = k; j = j+1;)
           DO (i = j; i = k; i = i+2;)
              b(j) = b(j)+a(i)*cf(i,j);
            CONTINUE comment: b coefficients;endcomment
          c(j) = b(j)*temd; temd = temd/d;
      CONTINUE
    IF (xm = 0)
       RETURN (a(),b(),c(),ff(),r())
     ELSE comment: can be used as BINOM.PSD; endcomment
      DO (i = 0; i = k; i = i+1;)
         bin(i,0) = 1;
          DO (j = 0; j = i; j = j+1;)
             bin(i,j) = bin(i,(j-1))*(i-j+1)/j;
```

```
CONTINUE
   CONTINUE comment: binomial coeffs;
         endcomment
  DO (i = 0; i = k; i = i+1;)
    m = k; l = k-i;
     temx = c(m)*bin(m,l);
     DO (j = l; j = 1; j = j-1;)
    m = m-1;
         temx = temx*(-xm)+c(m)*bin(m,(j-l);
     CONTINUE
    c(i) = temx;
  CONTINUE comment: c
             coefficients;endcomment
RETURN (a(),b(),c(),ff(),r())
```

---

**Table 5.09** Comparison of polynomial fits to the Normal density function
(a) $x$ range from –3.375 to 3.375 fitted at intervals of 0.375
(b) $x$ range from 0 to 4.5 fitted at intervals of 0.25

| *x* | *Calculated y* | *y(a)* | *y(b)* |
|---|---|---|---|
| –0.5 | 0.352065 | 0.353176 | 0.382026* |
| 0 | 0.398942 | 0.396392 | 0.398953 |
| 0.5 | 0.352065 | 0.353176 | 0.352157 |
| 1 | 0.241970 | 0.245561 | 0.241896 |
| 1.5 | 0.129517 | 0.128423 | 0.129582 |
| 2 | 0.053991 | 0.051251 | 0.053958 |
| 2.5 | 0.017528 | 0.020596 | 0.017515 |
| 3 | 0.004432 | 0.002099 | 0.004480 |
| 3.5 | 0.000873 | 0.019076* | 0.000815 |
| 4 | 0.000134 | 0.422661* | 0.000207 |
| 4.5 | 0.000016 | 2.42718* | 0.000033* |

Those values marked with * show how the fit deteriorates when the limits of the range are approached or exceeded.

---

With such a complex curve, all ten coefficients are required to give a reasonable fit. On the other hand, tan($x$) can be fitted to better than 1 ppm over the range 0 to 0.9.

Note; it is usually more accurate to use the *b*() coeficients with $z = (x-xm)/d$ for the determination of numerical values because the *c*() coefficients are subject to rounding error resulting from the rather complex nature of their computation. The *c*() expression should be reserved for investigation and clarification purposes.

## 5.03.5 Fourier Series

The representation of a function by a series of sine and cosine terms finds its major use in communications engineering, but it can also be useful in other fields. The technique is well–known and therefore this section will be limited to an algorithm and some examples to demonstrate the advantages and disadvantages of the method.

Over the range $x = -\pi$ to $\pi$, a function $y = f(x)$ may be represented by:

$$y = a(0) + \Sigma\, a(i)*\cos(i*x) + \Sigma\, b(i)*\sin(i*x) \quad (i = 1 \text{ to } \infty) \qquad 5.68$$

$$\text{where } a(0) = (1/2*\pi)*\int_{-\pi}^{\pi} f(x)*dx$$

and

$$a(\text{i}) = (1/\pi)*\int_{-\pi}^{\pi} f(x)*\cos(i*x)*dx$$

$$b(\text{i}) = (1/\pi)*\int_{-\pi}^{\pi} f(x)*\sin(i*x)*dx$$

Except for simple functions, the integrations can become quite tedious and, in most cases, it is only necessary to determine the order of magnitude of the $a$ and $b$ coefficients. Therefore, the accompanying subroutine and its BASIC equivalent use an approximate integration technique based on Simpson's rule which is described in Chapter 6.

Although the integrations in 5.68 appear, at first sight, to be quite innocent, they contain a nasty trap for the unwary user of approximate methods. As the index i increases, the gap between successive zeros of the sine and cosine functions decreases in proportion. Thus, if a fixed x interval for the determination of the ordinates is chosen, the approximate integration becomes progressively inaccurate as $i$ becomes larger and, if the number of intervals in a given half cycle is not an even integer, the contributions of the $y*\cos(i*x)$ and the $y*\sin(i*x)$ terms will not be properly balanced. These effects can lead to excessive errors if they are not taken into consideration.

The technique used in the algorithm for avoiding this difficulty is not the most elegant, but it is simple and it works. It is explained in the comments of the program. Here it is important to remember that it is not possible with this program to represent a series of isolated values by a Fourier series. The parent expression must contain continuous sections at least as long as the gap between the zeros of the highest order component used. Even so, the program does not perform too well on a simple square wave representation, because of the difficulty of accurately defining the vertical discontinuities.

SUBROUTINE FOURIER.PSD (nh,nc,FUNC,SIMPF,a(nh),b(nh))

comment: this program is restricted to an x range of $-\pi$ to $\pi$; for different ranges the appropriate conversion must be made. For example, let the original abcissa be u and the range from $u = a$ to $u = b$. Then we have:

$$x = 2*\pi*(u-a)/(b-a) - \pi \quad 5.69$$

and

$$u = (x+\pi)*(b-a)/(2*\pi) + a \quad 5.70$$

The values of *nh* (the order of the highest harmonic) and *nc*+1 (the initial number of ordinates) must first be selected. For reasons explained in Chapter 7, the normal number of ordinates for Simpson's Rule integration in this book has been chosen as 19, but the user is free to select any convenient odd number which gives satisfactory accuracy.

(For the examples *nh* = 10 and *nc* = 16 ).

The expression to be represented is assumed to be defined by a subroutine or function FUNC. This relation may contain finite discontinuities, but no infinite values. It is adviseable to try to arrange discontinuities between ordinates, if possible, in order to avoid errors. (See Exercise 5.4). The *x* range is then divided into *nc* equal intervals (*nc*+1 points) and the corresponding *y* values determined from FUNC, using 5.70 if necessary.

Using Simpson's Rule the expression represented by FUNC is then integrated over the x range at the interval $2*\pi/nc$ and the constant term found by dividing by the range.

The *a*(*i*) and *b*(*i*) are then determined from the appropriate relation in 5.68 with the correct number of intervals to avoid the loss of accuracy which would otherwise occur. The spacing must be chosen so that there is always an exact even number of intervals in each half cycle. endcomment.

```
CONST: INT nh,nc;   REAL pi;
VAR:   INT i,jf,ns;  REAL FUNC,SIMPF,h,xf,rnge;
                    REAL yy,a(nh),b(nh);

      BEGIN
        pi = 3.14159......;
        rnge = 2*pi;
        xf = -pi; jf = 0;
        SIMPF(nc,jf,rnge,FUNC,sumf);
        a(0) = sumf/2;
```

```
     DO (i = 1; i = nh; i = i+1;)
         jf = 1;
         ns = j*nc;
          SIMPF(ns,jf,rnge,FUNC,sumf);
          a(i) = sumf; jf = 2;
          SIMPF(ns,jf,rnge,FUNC,sumf);
          b(i) = sumf;
      CONTINUE
   RETURN (a(),b())
 END

SUBROUTINE SIMPF.PSD (ns,jf,rnge,FUNC,sumf)

 CONST: INT ns,jf;    REAL pi,fac,rnge;
 VAR:   INT i,nm;     REAL h,xf,FUNC,smp,sumf;

 BEGIN
   xf = -pi; nm = ns/2;
   sumf = 0; h = 2/3/ns
   DO (i = 0; i = ns; i = i+1;)
       IF (i = 0 OR i = ns)
           smp = 1;
       ELSE
         IF (i MOD (2) = 0)
           smp = 2;
         ELSE
           smp = 4;
         ENDIF
       ENDIF
    xf = rnge*(i-nm)/ns;
      IF (jf = 0)
        fac = 1;
      ELSE
        IF (jf = 1)
          fac = cos(j*xf);
        ELSE
          fac = sin(j*xf);
        ENDIF
      ENDIF
    sumf = sumf+smp*FUNC(xf)*fac;
   CONTINUE
    sumf = sumf*h;
   RETURN (sumf)
 END
```

### 5.03.5.1 Examples of Fourier analysis

The following examples were all obtained using the algorithm given above.

a. Square wave between 1 and –1

| *Harmonic* | *cos coeff* | *sin coeff* |
|---|---|---|
| 0 | 0 | – |
| 1 | 0 | 1.273 |
| 2 | 0 | 0 |
| 3 | 0 | 0.425 |
| 4 | 0 | 0 |
| 5 | 0 | 0.255 |
| 6 | 0 | 0 |
| 7 | 0 | 0.182 |
| 8 | 0 | 0 |
| 9 | 0 | 0.142 |
| 10 | 0 | 0 |

b. Sine wave with positive half cycle limited at 60°

| *Harmonic* | *cos coeff* | *sin coeff* |
|---|---|---|
| 0 | –0.013 | – |
| 1 | 0 | 0.971 |
| 2 | 0.027 | 0 |
| 3 | 0 | 0.023 |
| 4 | –0.019 | 0 |
| 5 | 0 | –0.014 |
| 6 | 0.009 | 0 |
| 7 | 0 | 0.005 |
| 8 | –0.002 | 0 |
| 9 | 0 | –0.001 |
| 10 | –0.002 | 0 |

c. Sine wave top and bottom limited at 60°

| *Harmonic* | *cos coeff* | *sin coeff* |
|---|---|---|
| 0 | 0 | – |
| 1 | 0 | 0.942 |
| 2 | 0 | 0 |
| 3 | 0 | 0.046 |
| 4 | 0 | 0 |
| 5 | 0 | –0.027 |
| 6 | 0 | 0 |
| 7 | 0 | 0.010 |
| 8 | 0 | 0 |
| 9 | 0 | 0.002 |
| 10 | 0 | 0 |

d. Parabolic rectifier with sine wave input: $y = x$^2; sine wave amplitude = 1; mid pt at $x = 2$

| Harmonic | cos coeff | sin coeff |
|---|---|---|
| 0 | 4.5 | – |
| 1 | 0 | 4.000 |
| 2 | –0.500 | 0 |
| 3 | 0 | 0 |
| 4 | 0 | 0 |
| 5 | 0 | 0 |
| 6 | 0 | 0 |
| 7 | 0 | 0 |
| 8 | 0 | 0 |
| 9 | 0 | 0 |
| 10 | 0 | 0 |

Note: (2 + sin*x*)^2 = 4 + 4*sin(*x*) + 0.5 – cos(2**x*)/2

## 5.03.6 Representation using well–known functions

The configuration of a given experiment or process can often give a clue to the possible characteristics of the resulting data and, although a good fit may be obtained with a polynomial, for example, such a curve does not necessarily correspond to the physical basis of the process. If the data are not contaminated with excessive noise, it is useful to examine the successive differences in order to establish whether a polynomial should be used or not. If the *j*th order difference tends to a constant, then the data may be satisfactorily represented by a *j*th order polynomial. On the other hand, if the ratio of the first differences in a given range seem to be constant, then a simple exponential term will probably be satisfactory over that range.

In section 5.02.4 above, it was shown that, although a given curve may be closely approximated by another function, the corresponding slopes can be very different. For piecewise replacement this can lead to serious problems at the change– over points.

A useful method of ensuring that the changeover slopes are similar is to fit a simple polynomial between two consecutive sections such that the polynomial overlaps both parts. In most cases, the 9th order fit with 19 data points is not necessary and a maximum of a 6th order fit with 7 points is usually close enough. The population curve of section 5.02.4 is again used as an example. Between $t = 0$ and $t = 500$, three separate expressions are used as approximations to the original. From $t = 0$ to $t = 250$, the relation *y* = 0.5*exp(0.01**t*) can be used, from $t = 250$ to 475, the cubic *y* = 14.2668 + 6.1169**z* + 1.0929**z***z* + 0.0431**z***z***z*; *z*=(t–350)/50 gives a reasonable fit and from $t = 475$ to 500 the straight line *y* = 0.25*(*t* – 529.33) + 50 is very close.

A rather coarse fit has been chosen deliberately to show the change over points clearly and the three approximations together with the parent curve are plotted in Fig. 5.05. Table 5.10 gives a comparison between the values computed from the equation of the curve and those derived from the approximations. The process can be repeated as often as is considered necessary to cover all parts of the data set. To obtain a better set of approximations, either choose a higher order polynomial with more data points or, alternatively, use smaller intervals if the data permits this.

---

**Table 5.10** Values of $F = 100/(1 + 199*\exp(-0.01*t))$ % for increments of 50 units in t with the relevant approximations:

$F1 = 0.5*\exp(0.01*t)$;

$F2 = 14.2668 + 6.1169 * z + 1.0929 * z * z + 0.0431*z*z*z$;

$(z=(t-350)/50)$

$F3 = 0.25*(t - 529.33) + 50$;

| *t units* | *F%* | *F1%* | *F2%* | *F3%* | *Least Error %* |
|---|---|---|---|---|---|
| 0 | 0.5 | 0.5 | | | 0 |
| 50 | 0.823 | 0.824 | | | –0.1 |
| 100 | 1.348 | 1.359 | | | –0.8 |
| 150 | 2.203 | 2.241 | | | –1.7 |
| 200 | 3.580 | 3.695 | 4.206 | | –3.2 |
| 250 | 5.769 | 6.091 | 5.954 | | –3.2 |
| 300 | 9.168 | 10.043 | 9.191 | | –0.3 |
| 350 | 14.267 | | 14.267 | | 0 |
| 400 | 21.529 | | 21.51 | | 0.1 |
| 450 | 31.146 | | 30.96 | 30.168 | 0.6 |
| 500 | 42.719 | | | 42.668 | 0.12 |

---

Values in this table should be compared with those in Table 5.05

The following rules should also be taken into consideration when making approximations to a curve.

a) Near a turning point ($dy/dx = 0$) the curve must have the form of a parabola, and,

b) Near a point of inflection ($d^2y/dx^2=0$) the curve must be nearly linear.

In general, there is no foolproof set of instructions that will cover all possible cases, but the best technique is to plot the data set first to obtain a picture of the form of the relation, and then polynomials can be fitted to selected parts of the set. If the resulting expressions have significant coefficients for powers higher than 5, then it is quite possible that transcendental functions are involved. When this is the case, the physical basis of the derivation of the data should be examined for clues as to the probable nature

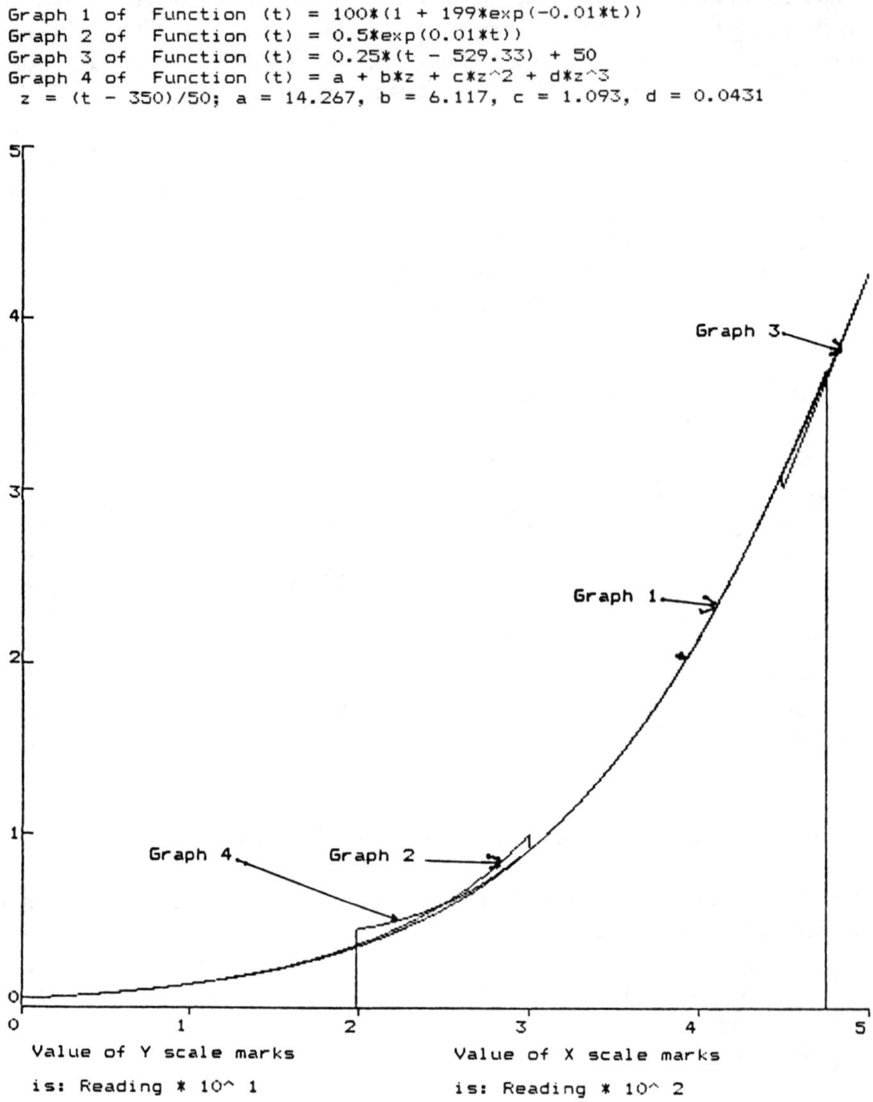
Graph 1 of Function (t) = 100*(1 + 199*exp(-0.01*t))
Graph 2 of Function (t) = 0.5*exp(0.01*t))
Graph 3 of Function (t) = 0.25*(t - 529.33) + 50
Graph 4 of Function (t) = a + b*z + c*z^2 + d*z^3
z = (t - 350)/50; a = 14.267, b = 6.117, c = 1.093, d = 0.0431
Graph 3
Graph 1
Graph 4
Graph 2
Value of Y scale marks
is: Reading * 10^ 1
Value of X scale marks
is: Reading * 10^ 2

**Fig. 5.05** Population curve and partial fit

of the connection between the variables. Examination of the polynomial coefficients may then indicate the type of transcendental to be expected. For example, if the terms are alternating in sign and only odd powers are significant, then a sine function might fit.

It is also advisable to try to determine whether there seems to be any asymptotes to the data points as this effect can derive from rational functions, from inverse relations, from expressions like $(a - b(\exp(-c*x))$ and from tangent functions.

It is usually not worth the effort to attempt to fit sums of transcendental functions unless it is previously known that the curve is of this form. Then sophisticated trial and error, in which parameters are varied by small amounts in a controlled fashion using an appropriate program, usually gives the quickest answer.

# Exercises 5

1. a) Show, using approximation techniques and without machine calculation, that, if $-1 < x < 1$, then

   sinh^2($x$) – sin^2($x$) ≈ ($x$^4)*2/3

   with an error of less than 1% over the whole range.

   b) Hence compute sinh^2(0.3) – sin^2(0.3) and estimate the fractional error.

   c) Compare the result of b) with direct machine calculation.

2. A wire of density 8000 kg/cu m with a yield stress of 5E8 N/m$^2$ is stretched between rigid supports 80 m apart. Show that if the sag at the centre of the wire is less than 0.12 m the wire will break under its own weight.

3. A square wave function is defined by the following relations,

   $y = -1; \quad -\pi < x < 0$
   $y = 1; \quad 0 < x < \pi$

   When these relations are used in the FOURIER algorithm of section 5.03.5 a false result is obtained. Explain and give the correct definition.

4. A glass tube with an outside diameter of 0.01 m and a wall thickness of 1 mm is drawn down until the length is ten times the original.

   Show that the new wall thickness is apoproximately $1/\sqrt{10}$ mm.

5. Given the following data set:

| | | | | | | | |
|---|---|---|---|---|---|---|---|
| $x$ | 0, | 0.05, | 0.1, | 0.15, | 0.2, | 0.25, | 0.3, |
| $y$ | 0, | 0.002459, | 0.0243, | 0.087070, | 0.2048, | 0.38068, | 0.60627, |
| $x$ | 0.35, | 0.4, | 0.45, | 0.5, | 0.55, | 0.6, | |
| $y$ | 0.86402, | 1.1287, | 1.3714, | 1.5625, | 1.6761, | 1.6931, | |
| $x$ | 0.65, | 0.7, | 0.75, | 0.8, | 0.85, | 0.9 | |
| $y$ | 1.6046, | 1.4147, | 1.1417, | 0.8192, | 0.4934, | 0.2187 | |

a) Plot the points
b) Fit a polynomial to the data
c) Deduce from the polynomial the $x$ and $y$ values of the maximum
d) Predict from the original plot the $y$ value at $x = 1$ and compare this value with that predicted by the polynomial.

6. a) Using the straight line equation: $y = x/\pi$ determine the values of the Fourier coefficients for the harmonic contributions necessary to provide a saw–tooth waveform of this shape.

b) The power in each of the components is proportional to the square of the corresponding coefficient. Estimate the mean square value of the linear wave form and hence determine the percentage of the total power in the fundamental.

c) Compare the sum of the squares of the components with the value of the mean square found in b) and estimate the percentage of the power not accounted for by the calculated harmonics.

# 6

# FUNCTIONAL APPROXIMATIONS: PART 3

## 6.01 Introduction

The techniques introduced in Chapters 4 and 5 dealt essentially with the replacement of functions and data sets by other, more easily determined relations. Dependent and derived relations and consequences of the replacements were considered only briefly.

In this chapter, the behaviour of functions near limits and asymptotes are considered. Approximate methods for evaluation of some types of integrals and differentials are presented with the limits of their applicability. A method of solving non–linear equations is given and finally some techniques for determining the effects of small changes in one or more variables in an expression.

## 6.02 Discontinuities and cusps

Discontinuities of differing types can occur quite frequently in engineering applications and special methods may be required in order to avoid system difficulties. For example, in a latitude– longitude coordinate system, an aeroplane flying over the north polar axis on a great circle route experiences a change of direction from north to south, although the angular velocity as measured from the centre of the earth has not altered and there is no change in the path of the machine. The variation is purely a consequence of the way in which the position and velocity of an object is described.

Question 3 of Exercises 5 illustrated a common difficulty in translating certain types of mathematical descriptions into a suitable form for machine calculation. The function was given as:

$$y = -1;\ (-\pi < x < 0)$$
$$y = 1;\ \ (0 < x < \pi) \qquad 6.01$$

Theoretically, the function is not defined at $x = 0$, but, in any practical situation there must be a mechanism for changing y from $-1$ to 1 and y must therefore be zero somewhere between $x = 0-$ and $x = 0+$. The ordinates in the subroutine FOURIER are spaced at intervals which depend on the order of the harmonic component. Therefore, as $x$ moves from negative to positive values, if $y$ is not defined at $x = 0$, the earlier value will be retained and the resulting wave form loses its symmetry leading to the generation of spurious components. This effect is a further consequence of the discontinuous nature of machine computation mentioned in Chapter 2, section 2.01. In fact, wherever a discontinuity occurs in a calculation, special precautions must be taken to ensure that the final answer is not seriously in error.

In equation 5.51 a series for $\tan(x)$ was derived. As $x$ tends to $\pi/2$, $\tan(x)$ tends to $\infty$, but $\tan(x) = 1/\tan(\pi/2 - x)$. Therefore if $u = \pi/2 - x$, we can write:

$$\tan(x) = 1/(u + .33333*u\text{^}3 + ......) \qquad 6.02$$

and as u becomes smaller $\tan(x)$ tends to $1/u$. This relationship gives a simple method for estimating tan(x) near $x = \pi/2$. Similar expressions hold for sec(x) and cosec(x) near $\pi/2$ and 0 respectively. For increased accuracy, the equation:

$$\tan(x) = \cos(u)/\sin(u) \qquad 6.03$$

may be used and the expression expanded using DIVSER and the binomial expansion. (See Exercises 6, question 1).

A curve has a cusp at a point $x = a$, if, when $x = a$, the two parts of the curve as $x$ approaches a from either side have a common tangent. For example the expression:

$$y\text{^}2 = x\text{^}3 \qquad 6.04$$

has a cusp at $x = 0$ where the tangent is parallel to the $x$ axis, whereas in the equation:

$$y\text{^}3 = x\text{^}2 \qquad 6.05$$

the cusp is such that the tangent is parallel to the x axis.

In all the above cases approximation techniques are complicated by the necessity of considering each section of the relationships separately in order to prevent mistakes. If such expressions occur as a result of manipulations during a program, they may be quite difficult to detect and it is therefore good practice to program in such a manner that tests can be made on each part independently to show up these possibilities.

In using expressions like 6.04 it should be noted that, in a program, any function can have only one result. The curve represented by $y\text{^}2 = x\text{^}3$, has, however, two quite distinct branches given by:

$$y = x^{1.5}; \text{ and } y = -x^{1.5}; \qquad 6.06$$

The second part of 6.06 must be interpreted as $-(x^{1.5})$ if the curve is restricted to the real domain. As $(-x)^{1.5}$ is an imaginary number, 6.06 has real solutions only for $x \geq 0$ and if $x$ becomes negative during a calculation, the program must either stop or change the sign in order to avoid an error.

These characteristics introduce problems in making approximations close to discontinuities and, in general, it is necessary to examine the data very thoroughly to find the discontinuity and make separate approximations on each side of the break.

# 6.03 Transformations of coordinate systems

In plane polar coordinates, the circle centred at the origin has the simple equation:

$r = R$; where $R$ is the radius of the circle
$r$ is the radius vector 6.07

In Cartesian coordinates the circle is:

$$x^2 + y^2 = R^2 \qquad 6.08$$

If 6.08 is written as $y = (R^2 - x^2)^{0.5}$, we again have the problem that the complete circle must be represented by two separate functions, and approximations near $x = R$ or $x = -R$ become more complicated.

The situation is improved by using the parametric form of the equations:

$$x = R*\cos(ang); \; y = R*\sin(ang); \; \tan(ang) = y/x; \qquad 6.09$$

and the expansions for cos and sin enable values for $x$ and $y$ to be determined easily.

In plotting routines for printers and $x$-$y$ plotters the programs for curves like the circle, the ellipse and similar functions, must use either the parametric equations or generate the separate parts of the curves by distinct methods. Many routines store the whole curve in an array with an entry for each 8 dots. For a picture of 420*540 dots, this technique requires a storage capacity of 25 kbyte for the array alone although most of the values will be 0. The storage requirement can be drastically reduced by generating and printing successively although the speed may also be reduced. For repeated plots of the same curve it is advisable to store the picture on disk.

## 6.03.1 A tyre must slide as it rotates

The following example illustrates the use of the parametric form in providing an approximate solution to a problem in which curved and linear functions are mixed.

Examination of a vehicle tyre shows that there is a flat portion in contact with the road surface. The circumference must remain very nearly constant, however, in order that the odometer and speedometer shall not give false results. If we assume that the part of the tyre not in contact with the road forms a circular arc, then the radius of this arc (*R1*) must be greater than the nominal tyre radius (*R*).

Let the length of the flat portion be 2**h*
the angle subtended at the axle be 2*alp

then the circumference of the tyre *C* is given by:

$$C = 2*\pi*R = 2*R1*(\pi - alp + \sin(alp)) \qquad 6.10$$

and

$$R1 = \pi*R/(\pi - alp + \sin(alp))$$
$$\approx \pi*R/(\pi-((alp)^\wedge 3)/6) \qquad 6.11$$

Thus in one revolution the material of the tyre moves a distance of $2*\pi*R1$ while the vehicle moves forward a distance of $2*\pi*R$.

The fractional difference *S* is:

$$S = 1/(1-((alp)^\wedge 3)/(6*\pi)) - 1$$
$$\approx ((alp)^\wedge 3)/(6*\pi) \qquad 6.12$$

which for an angle *alp* = 0.17 rad (10°) is about 0.025 %.

In a tyre life of, say, 50,000 km, the tyre has slid a distance of about 12.5 km, not including turns and skids, which, of course, do much more damage. When the tyre pressure is increased the amount of slip is decreased roughly as the cube of the pressure and this is one reason why tyres wear so much quicker when they are under-inflated.

Another way of looking at this effect is to consider velocities. The portion in contact with the road has a velocity *v* given by:

$$v = R1*\cos(ang)*d(ang)/dt \qquad 6.13$$

where ang is the angle between the vertical and the point at which the velocity is being taken. This may be considered as fixed point on the tyre, and therefore, as the angular velocity is constant, this fixed point has a variable velocity while it touches the road. The vehicle is, however, moving at a constant speed and there must be slip between the tyre and the road.

## 6.04 Limits and corners

Functions of the form $f(x)/g(x)$ can often tend to indeterminate expressions like 0/0 or $\infty/\infty$ for certain values of $x$.

Depending on the language and the machine being used, the program may crash, give an error message, or produce a false result. Many of these cases are dealt with in standard mathematical text books, but the judicious use of the relevant approximations can sometimes save a great deal of hunting. One important limit, x^n/n! was found in section 4.7 by such methods and further examples are given here.

1. $y = (1 - \exp(-k*x))/x;$ $x \longrightarrow 0$ 6.14

Expanding the numerator gives

$y = (1 - 1 + k*x + .....)/x$ 6.15

and, as $x \longrightarrow 0$, $y \rightarrow k$ 6.16

2. $y = x*\log(x),$ $x \longrightarrow 0$ 6.17

As $x \longrightarrow 0$, $\log(x) \longrightarrow -\infty$ and $y \longrightarrow -\infty * 0$ 6.18

$\log(x)$ can be written as $\log(1-(1-x))$ and, expanding, we have

$y = x*(-(1 - x) - (1 - x)\text{^}2/2 - ......)$ 6.19

Multipying out, it can be seen that, as $x \longrightarrow 0$, all the terms tend to 0,

Therefore, as $x \longrightarrow 0$, $y \longrightarrow 0$ also. 6.20

3. $y = 1/(1 + \exp(-1/x)),$ $x \longrightarrow 0$ 6.21

As $x \longrightarrow 0$ from the positive side, $y \longrightarrow 1$, but, if $x \longrightarrow 0$ from the negative side, $\exp(-1/x) \longrightarrow \infty$ and $y \longrightarrow 0$.

Expressions of this form are said to have a corner at the point of the discontinuity. There is no simple direct approximation for y at values of x close to 0.

4. $y = x^2 + x^2/(1 + x^2) + .... + x^2/(1 + x^2)^k +...; \quad x \to 0$ 6.22

This series is a geometric progression with a common ratio r where r = 1/(1+x^2), and the sum to infinity is:

$1 + x^2$ which tends to 1 as $x$ tends to 0 if $x^2$ is greater than 0, but if $x = 0$, all the terms are zero and the sum is then identically 0.

The situation is different, however if the number of terms is finite. Let $Sn$ be the sum of the n terms of the series in 6.22, then:

$$Sn = x^2*(1 - r^n)/(1 - r) \qquad 6.23$$

After some algebra and approximating for values of $x < 1$ we have:

$$Sn \approx n*x^2 \qquad 6.24$$

and this expression tends to 0 as x tends to 0 for all finite values of $n$.

In cases like examples 3 and 4 it is important to decide whether the practical application does or does not actually include the discontinuity and in 4 to determine whether the application really stems from a truly infinite series or just from a very large number of terms. Usually, it is found that processes are finite and essentially discontinuous and the approximations obtained by assuming values obtained by infinite sums can only be used when the relations do not contain pitfalls like those in examples 3 and 4.

# 6.05 Approximate Integration

## 6.05.1 Definite Integrals

A definite integral in plane coordinates can be regarded as an area between specified limits. In the case of rectangular axes the area is usually the area under the curve, whereas, in polar coordinates, the area is a sector between two radius vectors.

In Chapter 5, section 5.03.5, an approximate integration algorithm was used for the calculation of the Fourier harmonics of a wave form. This method was based on Simpson's rule which works extremely well for the great majority of applications. It is, however, essentially a cartesian process and there are cases in which a polar system can be simpler or give more reliable results. Two complementary subroutines are given with some examples to illustrate the difference between the polar and cartesian forms and so that, depending on the data to be processed, a choice can be made as to the more useful routine.

```
SUBROUTINE SIMP.PSD (ns,yy(),h,sumf)

comment: this algorithm uses a set of ns+1 data points as
y values in the array yy(ns). The number ns must be even.
The ordinates yy() correspond to x values at equal
intervals h. A summation is made according to Simpson's
rule and the result returned as sumf. endcomment

   CONST: INT ns; REAL h;
   VAR: INT i,nm; REAL yy(ns),smp,sumf;

   BEGIN
     sumf = 0;
     DO (i = 0; i = ns; i = i+1;)
         IF (i = 0 OR i = ns) smp = 1;
         ELSE
           IF (i MOD (2) = 0)
              smp = 2;
           ELSE
             smp = 4;
           ENDIF
         ENDIF
        sumf = sumf+smp*yy(i);
     CONTINUE sumf = sumf*h/3;
     RETURN (sumf);
   END

SUBROUTINE POLINT.PSD (ns,rr(),bet,sumf)

comment: this algorithm also uses a set of ns+1 data
points, but they are stored in the array rr(ns) as radius
vectors. The number ns is not restricted. Each radius
vector makes an angle bet with the preceding one where bet
is equal to the total angle between the limiting radius
vectors divided by ns.

The curve is then approximated by the arc of an
equiangular spiral for each sector, and the area in each
sector is then given approximately by:
```

```
Sector area = bet*(rr(i)^2+rr(i)*rr(i+1)+rr(i+1)^2)/6 6.25

the individual areas are then summed to give the total
area. endcomment.

      CONST: INT ns,; REAL bet;
      VAR: INT i,; REAL rr(ns),sumf;

      BEGIN
        sumf = rr(0)*(rr(0)+rr(1))+rr(ns)*rr(ns);
        DO (i = 1; i = ns-1; i = i+1;)
           sumf = sumf+(rr(i)*(2*rr(i)+rr(i+1));
        CONTINUE
           sumf = sumf*bet/6
        RETURN (sumf);
      END
```

Table 6.01 shows the results obtained for a selection of areas found using one or both of the algorithms for equations which can be integrated. Table 6.02 gives some values for expressions which can only be integrated numerically.

---

**Table 6.01** Integrable functions

This table shows some areas found using SIMP and POLINT. For some expressions it is possible to transform from one coordinate system to another without difficulty, and, where this is so, results have been given for both programs. In general, a polar system is more applicable for closed curves, whereas the rectangular coordinates are useful for relations which have no limits. All the curves shown can be integrated directly to obtain the true values for comparison purposes. The algorithms are, of course, most useful when mathematical integration is difficult, or impossible. (See Table 6.02).

Because of the form of POLINT, it may require more intervals to achieve a comparable precision than SIMP, but this is somewhat dependent on the nature of the relation being integrated. For example, POLINT gives a considerably more accurate result for the circle and ellipse.

It should be noted that where built–in machine functions such as square root are used, the error of the integration will be affected by the accuracy of the function.

The name of the curve is given in the second column and the corresponding equation in the appended auxiliary table.

All calculations are made with ns = 18.

| No. | Name | Limits | Type | Calculated Area | True Area | % Error |
|---|---|---|---|---|---|---|
| 1 | Straight line | $x = 0$<br>$x = 3$ | cart (SIMP) | 3 | 3 | Machine limits |
| 2 | Straight line | $ang = 0$<br>$ang = \pi/2$ | polar (POLINT) | 3.008 | 3 | 0.27 |
| 3 | Semi circle | $x = -1$<br>$x = 1$ | cart (SIMP) | 1.56226 | $\pi/2$ | 0.54 |
| 4 | Semi circle | $ang = 0$<br>$ang = \pi$ | polar (POLINT) | 1.57079 | $\pi/2$ | Machine limits |
| 5 | Semi ellipse | $x = -4/3$<br>$x = 4/3$ | cart (SIMP) | 2.40524 | 2.41839 | 0.54 |
| 6 | Semi ellipse | $ang = 0$<br>$ang = \pi$ | polar (POLINT) | 2.41636 | 2.41839 | 0.08 |
| 7 | Hyperbola | $x = -1$<br>$x = 1$ | cart (SIMP) | 0.255293 | 0.255673 | 0.15 |
| 8 | Hyperbola | $ang = 0$<br>$ang = \pi/2$ | polar (POLINT) | 0.256344 | 0.255673 | –0.26 |
| 9 | Sine curve | $x = 0$<br>$x = \pi$ | cart (SIMP) | 2 | 2 | Machine limits |
| 10 | Petal curve (sin(2**ang*)) | $ang = 0$<br>$ang = \pi/2$ | polar (POLINT) | 0.39071 | 0.39270 | 0.51 |
| 11 | Petal curve (sin^3) | $ang = 0$<br>$ang = \pi/2$ | polar (POLINT) | 0.244877 | 0.245437 | 0.23 |

| No. | Name | Equation: cartesian | Equation: polar |
|---|---|---|---|
| 1,2 | Straight line | *y* = –2/3**x*+2; | *r* = 6/(2*cos(*ang*)+3*sin(*ang*)) |
| 3,4 | Semi circle | *x*^2+*y*^2 = 1; | *r* =1 |
| 5,6 | Semi ellipse | 9*x^2/16 + 3**y*^2/4 = 1; | *r* = 1/(1–cos(*ang*)/2) |
| 7,8 | Hyperbola | *x*^2/0.64 – *y*^2/0.8 = 1; | *r* = 1/(1.5*cos(*ang*)+1) |
| 9 | Sine curve | *y* = sin(*x*); | – |
| 10 | Petal curve (sin(2**ang*)) | –; | *r* = sin(2**ang*) |
| 11 | Petal curve (sin^3) | –; | *r* = sin^3(*ang*) |

**Table 6.02** Non-integrable functions

This table shows some areas found using SIMP for some expressions which cannot be directly integrated. To demonstrate the convergence of the methods, the results have been calculated using two values of ns (4 and 18).

| *No.* | *Name* | *Limits* | *Type* | *Calculated area* $ns = 4$ | $ns = 18$ |
|---|---|---|---|---|---|
| 1 | Normal prob. density | $x = 0$<br>$x = 3$ | cart (SIMP) | 0.498479 | 0.49865 |
| 2 | Elliptic integral | $ang = 0$<br>$ang = 1$ | polar (SIMP) | 1.03733 | 1.03735 |
| 3 | Beta function | $x = 0$<br>$x = 0.5$ | cart (SIMP) | 0.67301 | 0.669931 |
| 4 | Gamma function | $x = 0$<br>$x = 5$ | cart (SIMP) | 2.77994 | 2.79753 |

| *No.* | *Name* | *Equation* | |
|---|---|---|---|
| 1 | Normal prob. density | y = 1/√(2*π)*exp(–x^2/2) | |
| 2 | Elliptic integral | y = 1/(1–0.25*sin^2(ang))^0.5 | Note: SIMP must be used as this is not an area |
| 3 | Beta(2.5,3.5) function | y = 27.167*x^1.5*(1–x)^2.5 | |
| 4 | Gamma (5) function | y = 1/24*exp(–x)*x^4 | |

## 6.05.2 Indefinite Integrals

An indefinite integral may be regarded as a function of the independent variable and there are two distinct approaches that can be adopted to find the required result.

If a set of data points will suffice, then repeated use of one of the definite integration procedures given above for the available values of the variable will produce the corresponding integrals. It is usually more accurate to move the lower limit as well as the upper limit and add the new value to the previous one, rather than start from a fixed lower limit each time. Care must be taken, however, to avoid rounding error and it is therefore better to overlap to a certain extent.

When a relation is necessary and the given function is not integrable, then it is sometimes possible to use the technique of Chapter 4, section 4.5 3 where a term can be replaced by an almost equivalent, but simpler expression which makes the integration possible. Unfortunately, as the equations differ so much, it is not possible to give rules that will cover every case, but the following methods may be used as a guide.

a) If a term varies very slowly over the range, replace it by its average value.

b) When possible, expand the expression in a series and integrate term by term, making sure that the resulting series also converges to the correct result.

c) For limited ranges, individual terms may be replaced by their expansions and sufficiently small values neglected.

d) If a polynomial gives a satisfactory fit over a given range, then the integral of the polynomial will correspond to the required integral over that range.

**Examples**

1. $I = \int \exp(-x*x/2)*dx$ 6.26

Expanding this expression gives:

$$I = \int (1 - x*x/2 + x\text{^}4/4/2 - \ldots.)*dx \qquad 6.27$$

and $I = x - x\text{^}3/6 + x\text{^}5/40 - \ldots\ldots + \text{constant}$ 6.28

For $x < 1$, 6.28 can be used to determine $I$ as the constant is known to be $2*(\sqrt{}(\pi/2))$.

2. $I = \int (\log(x\text{^}2 + a\text{^}2))*dx$ 6.29

This expression is, in fact, integrable, but the result is so messy that, if $x/a < 1$, it is easier to use the approximation.

Expanding, we get:

$$I = \int(\log(a\text{^}2) + (x\text{^}2/a\text{^}2 - x\text{^}4/(2*a\text{^}4) + \ldots.))*dx \qquad 6.30$$

and

$$I = \mathrm{x}*(\log(a\text{^}2) + x\text{^}2/(3*a\text{^}2) - x\text{^}4/(10*a\text{^}4) + \ldots.) \qquad 6.31$$

3. The time of rise and growth of a vapour bubble in a liquid has important consequences in moulding and other fields, but the equations are complex. An approximate solution may be obtained as follows.

|  | | |
|---|---|---|
| Let | the surface tension of the liquid be | $T$ |
| | radius of the bubble be | $a$ |
| | depth of formation of the bubble be | $h$ |
| | height from this point be | $z$ |
| | atmospheric pressure be | $P0$ |
| | viscosity of the liquid be | $vc$ |
| | acceleration due to gravity be | $g$ |
| | density of the liquid be | $ld$ |
| | mass of vapour in bubble be | $m$ |
| | gas constant at the temperature be | $B$ |
| | total pressure at depth h be | $P$ |
| | vapour pressure inside the bubble be | $Pv$ |

then, assuming that the acceleration is so fast that the velocity of the bubble is effectively the terminal velocity we have:

$$dz/dt = 2/9*g*ld*a\text{^}2/vc \qquad 6.32$$

Observation shows that the radius increases as the bubble rises and it is therefore necessary to find a relation involving $a$ so that 6.32 can be solved, at least approximately. If we assume that the rate of the physical and chemical reactions in the process are much faster than the rate of rise of the bubble, then an equilibrium between the external pressure and the internal forces is approximately achieved at all points on the path of the bubble. Thus we may write:

$$P - g*ld*z = 2*T/a + Pv \qquad 6.33$$

and, solving for a:

$$a = 2*T/(P - Pv - g*ld*z) \qquad 6.34$$

Substituting for a in 6.32 and rearranging we get:

$$(P - Pv - g*ld*z)\text{^}(2)*dz = C*dt \qquad 6.35$$

where $C = 2*g*ld*T*T*4/9/vc$ 6.36

The term $P - Pv$ represents the pressure difference between the inside and outside of the bubble at the starting point at depth $h$ below the surface. If we make the assumption that Pv stays approximately constant, 6.35 can be integrated directly.

Integrating and adding the constant of integration, we have:

$$(P - Pv)^{\wedge}(3) - (P - Pv - g*ld*z)^{\wedge}(3) = 3*g*ld*C*t \qquad 6.37$$

Let the diameter of the bubble at the surface, just before it bursts, be *bd*, then from 6.33

$$P - Pv - g*ld*h = T/bd \qquad 6.38$$

In the case of liquid metals $g*ld$ is of the order of 70 000 and bubbles as large as 1 cm diameter or more are of interest. The surface tension of most metals at the temperature of pouring is in the region of 1.5 N/m so that $P–pv$ is almost equal to $g*ld*h$ and 6.37 approaches a limiting form:

$$z*(3*h*h - 3*h*z + z*z) = 3*C*t/(g*ld*g*ld) \qquad 6.39$$

from which the time of rise may be estimated.

For lower density liquids and smaller bubbles, however, $P - Pv - g*ld*h$ is significant and must be determined from 6.38 before t can be calculated using 6.37. See Exercises 6 number 3.

# 6.06 Approximate differentiation

Because differentiation is essentially the ratio of two differences which approach zero together, there will inevitably be loss of significance when an attempt is made to determine a differential coefficient approximately from numerical data.

Except for a few exotic functions, nearly all the expressions that are likely to be met can be differentiated so that in this section we shall only be concerned with deriving differentials from data sets.

Consider the following data set:

| | *0* | *1* | 2 | *3* | *4* | *5* | *6* |
|---|---|---|---|---|---|---|---|
| $x$ | 1 | 1.1 | 1.2 | 1.3 | 1.4 | 1.5 | 1.6 |
| $y$ | 0.8415 | 0.8912 | 0.9320 | 0.9636 | 0.9854 | 0.9975 | 0.9996 |

and suppose that we need to estimate the slope at $x = 1.32$, $x = 1.02$ and $x = 1.58$. The difference between the $x$ values = $h$.

Three methods of different precision are given and the results are tabulated in Table 6.03.

## 6.06 1 Linear difference (1)

The slope is estimated between successive points by:

$$slope = (y(i + 1) - y(i))/h \tag{6.40}$$

and the value is assumed to be constant over the interval.

$$\begin{aligned} slope(1.32) &= (0.9854 - 0.9636)/0.1 \\ &= 0.2189 \end{aligned} \tag{6.41}$$

## 6.06.2 Linear difference (2)

The slope is estimated at a given point $y(i)$ by:

$$slope = (y(i + 1) - y(i - 1))/(2*h) \tag{6.42}$$

and the intermediate value is found by linear interpolaton between successive values.

## 6.06.3 Curve fitting

In this method, a polynomial (or other suitable curve) is fitted to the points and the required slope found by direct calculation of the differential at the specified point. The data set consists of seven points so that a 6th order polynomial can be fitted.

**Table 6.03** Slope estimation
The data set of 6.06 is taken from a table of sines rounded to four significant figures.

| *Method* | *x* | *Estimated Slope* | *Correct value* | *% Error* |
|---|---|---|---|---|
| 1 | 1.02 | 0.497 | 0.5234 | 5 |
| 2 | 1.02 | – | | – |
| 3 | 1.02 | 0.5233 | | 0.02 |
| 1 | 1.32 | 0.2189 | 0.2482 | 11 |
| 2 | 1.32 | 0.2475 | | 0.3 |
| 3 | 1.32 | 0.2479 | | 0.12 |
| 1 | 1.58 | 0.021 | –0.0092 | 328 |
| 2 | 1.58 | – | | – |
| 3 | 1.58 | 0.0097 | | –5 |

In many cases, method 2 is quite adequate, but method 1 is not recommended. Method three needs more time and effort, but for complex curves it is usually safer.

Note that none of the methods can be relied on near the ends of the range. If the application requires slope estimation at the limits and more data points cannot be determined, then method 3 should be used, but care must be taken in the use of the slopes calculated in this way.

# 6.07 Expressions involving small incremental angles

The differential $dy/dx$ at the point $(x,y)$ is defined as:

$$dy/dx = \text{Limit } (y1 - y)/(x1 - x) \text{ as } x1 \longrightarrow x \qquad 6.43$$

Let the value of the differential at $(x.y)$ be diff($x$)

Then in some cases, it is possible to write with sufficient precision:

$$y1 = y + \text{diff}(x)*(x1 - x) \qquad 6.44$$

However, if diff($x$) is varying rapidly with $x$, then this relation may not be good enough. When the Taylor's expansion is well-behaved, higher terms can be used, but when the expression involves trigonometrical functions another technique may be better.

## 6.07.1 Example 1

A simple example of a range finder with a base line ($b$) of 1 m illustrates the method. Let us suppose that one object makes an angle of *ang* rad at the end of the base line and a second object makes an angle of ***ang1***. Then, from 6.44, if *ang* = 1.50 and ***ang1*** = 1.53:

Distance between the objects = 1*sec^2(1.50)*(0.03) = 6 m 6.45

whereas a more accurate calculation gives:

Distance between the objects = 10.4 m 6.46

and, quite obviously, 6.45 cannot be accepted.

Therefore, let the distance of the objects be $r$ and $r1$ and the difference between the angles be $p$, then:

$r1 - r = b*\tan(ang + p) - b*\tan(ang)$

$$= b*((\tan(ang) + \tan(p))/(1 - \tan(ang)*\tan(p)) - \tan(ang)) \qquad 6.47$$

If we assume that $p$ is such that $\tan(p) = p$ then 6.47 can be simplified to:

$$r1 - r = r/(\cot(ang)/p - 1) \qquad 6.48$$

giving $r1 - r = 10.3$ m 6.49

which is much closer.

The technique is thus to expand the trigonometric functions of the composite angles and, where possible, use: $\sin(p) = p$, $\tan(p) = p$ and $\cos(p) = 1$. If the first order values tend to cancel, then it may be necessary to take the next terms of the expansions in order to achieve a satisfactory result.

## 6.07.2 Example 2

The next example demonstrates this effect. A radar system measures the position of an aeroplane in polar coordinates, $(r,el,az)$, where $r$ is the radius vector, $el$ is the angle of elevation and $az$ is the azimuth angle. A new position is calculated every 10 seconds. For traffic control purposes, the average speed of the aeroplane must be known and this can be found from the expression:

$$v = (r1\string^2 + r\string^2 - 2*r*r1 * \cos(el1 - el)*\cos(az1 - az))\string^0.5/10 \qquad 6.50$$

Put $r1 - r = s$, $el1 - el = p$ and $az1 - az = q$ 6.51

Now, when the machine is at a distance greater than about 30 km, the equation 6.50 tends to involve the difference of two nearly equal large values and angles $el1-el$ and $az1-az$ both become small so that the calculation loses precision due to measurement and rounding errors. Using 6.51, 6.50 becomes:

$$v = (s\string^2 + 2*r1*r*(1 - \cos(p)*\cos(q)))\string^0.5/10 \qquad 6.52$$

$$= (r\string^2*((s/r)\string^2 + (1 + s/r)*(p\string^2 + q\string^2)))\string^0.5/10 \qquad 6.53$$

giving $v \approx r/10*((1 + s/r)*(p\string^2 + q\string^2) + (s/r)\string^2)\string^0.5$ 6.54

If r = 40 km, $r1 = 40.15$ km, $p = 0.031$ rad and $q = 0.035$ rad

$v = 0.1879$ m/s = 676.7 km/hr from 6.54

and

$v = 676.6$ km/hr from 6.52

Using suitable equipment the values $r,s,p$ and $q$ can be measured directly for immediate entry into the computing system.

## 6.08 Approximate Solutions of Identities

A rapid evaluation of square and cube roots is often required and the following method gives very good results.

Let $x$ be the number and $y$ the square root, then:

$$y*y = x \qquad 6.55$$

Guess a value $y1$ for $y$ (usually the nearest easy root) and let

$$y = y1 + h \qquad 6.56$$

Then we have:

$$(y1 + h)*y1 + h) = x \text{ and } y1*y1 + 2*y1*h + h*h = x \qquad 6.57$$

Assuming $h*h$ may be neglected compared with $y1*y1$, 6.57 becomes:

$$h = (x - y1*y1)/(2*y1) \qquad 6.58$$

An improved value $y2$ can be formed by adding $h$ to $y1$ giving:

$$y2 = (x + y1*y1)/(2*y1) \qquad 6.59$$

This process can be repeated as often as necessary until the desired precision is obtained. An algorithm for square roots is given below. A similar technique can be used for higher roots and the derivation is left as an exercise (See Exercises 6 No. 4).

### 6.08.1 `SQROOT.PSD(x,y,eta)`

```
comment: the square root of the given value x is returned
as y to the precision eta. If required, eta can be the
machine precision and it can then be included
automatically. Note that, in this case, it is not
```

```
sufficient to put eta equal to zero because the difference
of y and yy can vary by one least significant bit due to
rounding error. In order to avoid the necessity of
including the first guess, y1 is taken as x/2 if x >1 and
2*x if x < 1. endcomment.

       CONST: REAL x,eta;
       VAR    REAL y,yy;

       BEGIN
         IF (x<1)
            y = 2*x;
         ELSE
            y = x/2;
         ENDIF
            yy = y;
          WHILE (abs(y-yy)>eta)
             y = (x + y*y)/2/y;
            yy = y;
          ENDWHILE
         RETURN (y)
       END
```

## 6.08.2 Example: Square root of 0.75

Take the initial $y$ as 1.5, then using 6.57, we get:

$y1$ = 1
$y2$ = 0.875
$y3$ = 0.86607
$y4$ = 0.8660268 = 0.8660 to four significant figures
$y5$ = 0.866025437
$y6$ = 0.866025403
$y7$ = 0.866025404
$y8$ = 0.866025404 to nine significant figures.

Squaring 0.8660 gives 0.749956 and $y8*y8$ = 0.75 to nine figures.

# 6.09 Approximate solutions of equations

## 6.09.1 Linear Equations and Ill–conditioning

The methods for the solution of systems of linear equations are very well known (Bib 3,4,11 etc.)and they will not be given here. However, there are some aspects which can be dealt with by approximation techniques.

Consider the following pair of equations:

$$y = m*x + c \; ; \qquad y = r*x + b \qquad 6.60$$

The solution of these for y is:

$$y = (r*c - m*b)/(r - m) \qquad 6.61$$

Now put $r = m + h$ and $b = c + g$

where $h/m$ and $g/c$ are both much less than 1, then:

$$y = m*x + c \; ; \qquad y = (m+h)*x + c + g \qquad 6.62$$

and $y = c - g*m/h$ 6.63

In principle, 6.61 and 6.63 should give the same result, but, in practice, 6.63 is to be preferred when $g$ and $h$ are small. When m and r are nearly equal the lines (or more dimensional figures in more complex sets of equations) are nearly parallel and the solution is very dependent on the value of h. Differentiating with respect to h, we have:

$$dy/dh = g*m/h*h \qquad 6.64$$

That is: the rate of change of $y$ with $h$ is inversely proportional to $h$^2 and if $h$ is small $dy/dh$ can be very large. When this is the case the equations are said to be ill–conditioned and high precision arithmetic is necessary to obtain accurate results.

For sets of linear equations the substitution of relations like 6.63 for 6.61 with the subsequent determination of $dy/dh$ etc., is worth while if ill–conditioning is suspected.

## 6.09.2 Iterative Improvement of Solutions

Rounding error can often spoil the accuracy of the solution of a set of linear equations, even when the all the coefficients are known to greater precision than machine accuracy. An improvement can often be obtained by using an approximation technique similar to, but more complex than, those previously described in section 6.08.

Although a treatment of matrix methods is beyond the scope of this book, we shall need the following notation in order to describe the technique.

A matrix is indicated by bold symbols (including the * for multiplication) and the expression:

$$\mathbf{A*x = b} \qquad 6.65$$

is to be interpreted as a set of linear equations connecting the matrix product of the matrix **A** with the column vector **x** and the constant column vector **b**.

Suppose that a solution of 6.65 **x0** has been found by one of the methods for solving such sets, then a column vector called a residual vector **r0** can be defined by:

$$\mathbf{r0 = b - A*x0} \qquad 6.66$$

and it can be shown that, if **X** is the solution of the auxiliary equation:

$$\mathbf{A*X = r0} \qquad 6.67$$

then an improved solution **x1** of 6.65 is given by:

$$\mathbf{x1 = x0 + X} \qquad 6.68$$

These relations 6.66 to 6.68 can be iterated, although rounding error will inevitably limit the precision of the result. In many cases just one iteration is sufficient.

It is *most important* that the residual vector **r0** should be calculated with higher precision than the other terms of the relation.

More information on matrix methods can be found in Press et al (Bibl 1), Westlake (Bibl 12) and Golub and van Loan (Bibl 13).

## 6.09.3 Non–linear Equations

The solution of a non–linear equation is rarely an exact integer and there are very few cases in which it is possible to find an algebraic answer. Several numerical methods of finding roots are available, but it is considered that the method of false position is preferable for most applications. It converges quickly and requires only one evaluation of the function for each iteration. The only major disadvantage is that two starting values, *x0* and *x1*, are required, one each side of the root. This difficulty is easily overcome by using an auxiliary algorithm STARTFP to determine the starting values prior to entry into the main program. An auxiliary program to locate a starting point is necessary for all methods, in any case. The loss of time caused by the determination of the second starting value is more than offset by the rapid convergence.

The method of false position can be explained as follows. If a solution $x$ of the equation y = f(x) lies between $x0$ and $x1$, then a first estimate of $x$ is given by:

$$x = x1 - (x1 - x0)*f(x1)/(f(x1) - f(x0)) \qquad 6.69$$

If f(x) is the same sign as f(x1), x0 is put equal to x1 and x1 equal to x, otherwise x0 = x and x1 stays unchanged. 6.69 is then revalued and the procedure repeated until the required precision is reached. The algorithm FALSPOS gives the code for the technique.

```
SUBROUTINE STARTFP.PSD(xs,h,n,x0,x1,FUNC)

comment: this routine calculates successive values of the
function FUNC beginning with xs and incrementing xs by h at
each step for n steps. When the product FUNC(x)*FUNC(x+h) is
negative, then there must be a root between x and x+h. If h is
small enough, so that there is only one root in this interval,
the values of x and x+h are returned as x0 and x1 for use in
FALSPOS. If there is no root in the chosen range, the procedure
must be repeated until one is found or the search for a root
abandoned. end comment.

      CONST:  INT n;  REAL xs,h;
      VAR   :  INT i;  REAL x,x0,x1,y0,y1,FUNC();
                       CHAR ans;
      BEGIN
       x0 = xs;
       x1 = xs + h;
       y0 = FUNC(x0);
       y1 = FUNC(x1);
        i = 0;
         WHILE(i<n)
            IF (y0*y1<0)
         WRITESCR("Do you want to try a smaller interval?")
              INPUT (ans)
                IF (ans = "y" OR ans = "Y")
                  h = h/2;
                ELSE
                  RETURN (x0,x1)
                ENDIF
            ELSE
```

```
                y0 = y1;
                x0 = x1;
                x1 = x1 + h;
                y1 = FUNC(x1);
              ENDIF
              IF (y0*y1>0 & i = n)
                WRITESCR("There is no root in this range")
                WRITESCR("Do you want to continue?)
               INPUT (ans)
                 IF (ans = "y" OR ans = "Y")
                   i = i+1;
                   n = n+n;
                 ELSE
                   x0 = 0;
                   x1 = 0;
                   RETURN (x0,x1)
                 ENDIF
              ELSE
                 i = i+1;
              ENDIF
            ENDWHILE
        END
```

SUBROUTINE FALSPOS.PSD(x,x0,x1,eta,FUNC)

comment: the routine takes two values of x (x0 and x1) between which a root is known to lie. Corresponding y values (y0 and y1) are calculated from the given equation FUNC(). A straight line is drawn between y0 and y1 and the intersection of this line with the x axis is then determined. This new value (x) is an estimate of the root which can be improved by iteration until two successive values differ by less than eta.

If required, the WHILE loop can include a counter to determine the number of iterations, but this is only necessary when system or function comparisons are being made.

Caution: with badly behaved functions like sin(1/x) as x tends to zero, the method can fail and, therefore, a protection is included to detect the failure. endcomment.

```
        CONST:  REAL eta;
        VAR  :  INT i;  REAL x,x0,x1,x2,y0,y1,z,z1,FUNC();
```

```
BEGIN
   y0 = FUNC(x0);
   y1 = FUNC(x1);
   x  = x0;
   x2 = x1;
   z  = abs(x2-x);
   z1 = z;
      WHILE (z > eta)
        IF (z > z1)
          WRITESCR ("System does not converge")
          WRITESCR ("Check function and start values")
          RETURN
        ENDIF
          z1 = z;
          x2 = x;
            IF (y0*y1<0)
              x0 = x;
            ELSE
              x0 = x1;
              x1 = x;
            ENDIF
          x = x1-(x1-x0)*y1/(y1-y0);
          z = abs(x2-x);
         y0 = FUNC(x);
            IF ((y1-y0) = 0)
              ENDWHILE
            ENDIF
      ENDWHILE
    RETURN (x)
  END
```

# Exercises 6

1. Given that $y = \sec(x)*\tan(x)$, where $x = \pi/2 - 0.01$, estimate $y$, using the three methods indicated, to a precision of six decimal digits and compare the answers with y = 9999.83, a value calculated directly from the expression with a basic precision of 10 digits, but rounded to 6 figures.

   a). $y = 1/(\sin(0.01)*\tan(0.01))$

   b). $y = \cos(0.01)/\sin(0.01)/\sin(0.01)$

   c). $y = \cos(0.01)/(1 - \cos(0.01)*\cos(0.01))$

2. A single–sided range finder has a nominal base line of 1 m. The mirror angle can be measured to a precision of 0.0002 rad and the base–line material has a temperature coefficient of 5E–6. The instrument is to be used over a temperature range of 40° C.

   a). What is the maximum distance that can be measured with an error less than 20 m ?

   b). What is the maximum error at a distance of 2 m ?

3. A bubble in a glass of champagne starts 10 cm below the surface. (For the flat type of champagne glass use 3 cm.)

   If the diameter of the bubble at the surface is 0.5 mm, calculate the time that it takes to reach the surface. Take the density as 970 kg/cu m, the viscosity as 1.3E–3 Ns/m$^2$ and the surface tension as 0.055 N/m.

   Note that as explained in section 6.05.2 example 3, $P - Pv - g*ld*z$ cannot be neglected in this case.

   Compare your results with experiment.

4. Following the method of section 6.08:

   a). Derive iterative methods of finding the cube and fourth roots of a number, including the means for selecting the starting points.

   b). Compare the results obtained for the cube root of 27.33 and the fourth root of 256.89 with those obtained by using the binomial expansion technique.

5. Fit a polynomial to the catenary: $y = 50*\cosh(x/50)$, between $x = -50$ and $x = 50$:

   a). With $y$ as the polynomial ordinate.

   b). With $y - c$ as the ordinate.

   c). With the same $x$ values repeat a and b for the catenary $y = 500*\cosh(x/500)$ and then compare the results.

   d). Find the slopes of all the curves at $x = 18.5$ using the methods of section 6.06.

6. Integrate the expression: $y = 2*\exp(-x*x)$ between $x = 0$ and $x = 4.5$ and hence estimate the square root of $\pi$.

a). Using SIMP

b). If the right-hand side is expanded, integrated term by term and summed, a false result may be obtained. Why ?

7. The equation: $y = \cos(x)/x - \sin(x)/x/x$ defines the amplitude of the resultant radiation after passage through a slit.

   a). Find the first five roots of the equation using FALSPOS, given that all the roots are close to $(2*n - 1)*\pi/2$, $(n = 1,2,3,...)$

   b). Solve it by expanding and approximating.

# 7

# STATISTICS

## 7.01 Introduction

There is no doubt that statistics is a difficult subject. Not only are the basic concepts rather unfamiliar, but the mathematical relations mostly involve transcendental functions which can only be dealt with by numerical methods or relatively complicated tables.

Certainly, the Normal (Gaussian) distribution is now well known, but there is a strong tendency to apply it to every case in which a result or problem is affected by random fluctuations. Unfortunately, as engineering systems become more complex and higher precision is required, it will be found that the standard formulas are no longer good enough.

The object of this chapter is to present techniques for making adequate approximations to some of the calculations that occur in processes subject to probabilistic variations. In order to avoid difficulties with names and methods, a few definitions and explanations are given, but the reader is assumed to be familiar with the general concepts and techniques of statistical methods. Some texts at different levels are given in the bibliography.

The *probability* (P) of an event is the ratio of the number of times the event occurs to the total number of trials in which that event is considered when the number of trials becomes very large.

A *random variable* is defined as a quantity, the value or values of which are not determined precisely by a given relationship, but which may subject to indeterminate fluctuations.

If, instead of a discrete number of trials, we are concerned with continuous variables, it is no longer possible to define a probability of a single point and this leads to the concept of a probability expressed as the product of a variable ($xP$) with an infinitesimal ($dx$) of the independent variable $x$.

The value $(xP)$ is known as a *probability density* and it has the dimensions of probability per unit $x$. When $xP$ is given as a function of $x$, we have a *probability density function.*

When the sum of all the probabilities up to a certain point is given as a function of the number of trials (discrete case) or the value of the independent variable at that point (continuous case), the function is known as the *probability distribution function*, or simply as the *distribution function* if there is no possibility of confusion. In the continuous case, of course, the sum becomes an integral.

It is very important to keep the distinction between distribution functions and density functions quite clear, as misunderstanding can lead to total chaos in situations where the interpretation is, to say the least, perhaps somewhat obscure.

One of the most remarkable conclusions of the study of probability is the *Central Limit Theorem* which states that the probability distribution function of the sum of a sufficient number of random variables will tend to the Normal distribution, whatever the distribution of the individual variables.

As many observed quantities are effectively due to the contributions of large numbers of more or less randomly varying factors, the ubiquity of the Normal distribution is not surprising. The snag comes, however, with the term "a sufficient number". Sometimes, as with simple selection problems, the number may be as low as five, but, in other cases where there are unknown causes at work, the distribution can be very far from Normal. Size distributions of a single material are rarely Normal although they can be considered to be the end product of complex processes,

In describing random distributions it is often convenient to have a few numbers which summarise the characteristics. There are four such values in common use, They are:

The *mean* $(mn)$ or average given by $mn = (\Sigma x)/n$ 7.01

The *median* $(med)$ which is that value having an equal number of outcomes above and below it. 7.02

The *mode* $(mx)$ which is the most frequently occurring outcome,

and the *standard deviation* (sd) given by:

$sd = \sqrt{(\Sigma(x-mn)^2/n)}$ for a complete population of $n$ values 7.03

or

$sd = \sqrt{(\Sigma(x-mn)^2/(k-1))}$ for a sample of size $k$ from a larger population 7.04

The *variance* $(vr)$ is the square of sd 7.05

The standard deviation is a measure of the spread of values of the variable about the mean, but its usefulness depends on the type of distribution.

Again the distinction between 7.03 and 7.04 must be clearly understood although it is only important for values of $n$ which are less than about 30. One useful way of remembering which is which is to consider a sample of 1 out of a population of $n$. Thus, although 7.03 can give a value for a population of 1, it is obvious that a sample of 1 can tell us nothing about how the values are distributed about the mean of $n$ components and, in that case, sd is not defined.

Approximations in statistical calculations are of two general kinds which have a rather subtle distinction:

a). Functional and numerical, as described in previous chapters.

b). Cases in which an experimentally determined value is substituted for the unknown quantity being sought.

The validity of the use of type b can only be substantiated after the calculation is complete and further experiments have been carried out to test the conclusions.

This aspect is typical of probability and statistical work. An answer can never be exact, but justifiable only within limits which are themselves subject to uncertainty.

One other point should be mentioned here. There is a tendency to regard low probabilities as being unimportant when they are of the order of 1E-6, but diamonds are found and lotteries are won even though the chances in both cases are much lower than 1E-6.

Each case must be considered on its merits, and when the number of experiments is large, even events with very small possibilities have appreciable chances of occurring.

# 7.02 Binomial Probability

An event is described by *binomial probability* when it can either happen (success) or not (failure). Such events are very common. Classical cases include, finding a particular type of animal in a large population, selection of the right number in a lottery, tossing a coin or living long enough to collect your endowment assurance. An actress was once asked if she were pregnant and the reply came, "only a little bit", but, sad to say, she was talking about a binomial random variable and the outcome was "success". This story highlights a difficulty often met with in binomial experiments, if the variable being examined is, say, the occurrence of aeroplane crashes, then any crash is called a success for the purpose of the study in spite of its macabre associations.

## 7.02.1 How many measurements ?

Procedures for estimating blood counts, atmospheric polluting particle concentrations, mineral grain abundance in an ore sample and traffic density, for example, are all methods which involve counting techniques for the subsequent determination of a ratio (number per unit of volume, area, time or other variable). These procedures, in spite of automatic methods, may be time consuming and expensive. It is therefore desirable to know, as soon as possible, when enough results have been accumulated to give the desired accuracy.

The subject of interest is usually the proportion of a given species in the mass and a success is defined as the finding of a particle of that species.

| Let | | |
|---|---|---|
| | the probability of success be | $P$ |
| | number of trials be | $n$ |
| | actual number of successes be | $s$ |
| | proportion of the species be | $F$ |
| | maxium permissible fractional error be | $fe$ |

then the probability of failure is $Q = 1 - P$ 7.06

and the mean number of successes is $mn = n*P$ 7.07

Therefore $F = mn/n = P$ 7.08

For a binomial variable the variance for n trials is given by:

$$vr(s) = n*P*Q = n*P*(1 - P) = mn*(1 - P) \qquad 7.09$$

and the variance for the proportion by:

$$vr(F) = P*Q/n \qquad 7.10$$

Now, if enough experiments have been made, we hope that the observed proportion s/n is close to P, so, substituting this quantity in 7.10 we get:

$$vr(F) \approx s/n*(1 - s/n)/n \qquad 7.11$$

Before doing the experiments, a decision must have been made as to the required precision of the answer and the allowable uncertainty of that precision.

For the binomial distribution, the calculations for the limits of the variance are very complex and to obtain a relatively simple answer, approximate methods must be used.

There are two possibilities: if $mn$ is known to be less than 5, with n large, the Poisson distribution is a good approximation to the binomial; but if $mn$ is greater than 5 and $P*(1 - P)$ is not less than 0.1, then the Normal distribution is applicable. Taking the second case first and replacing the binomial distribution with its Normal approximation with a mean of $n*P$ and sd of $\sqrt(vr)$, we can say that the standard error of the mean (*sem*) is given approximately by:

$$sem(s) = \mathrm{sd}/\sqrt(n) \qquad 7.12$$

or $$sem^2(F) = P*(1 - P)/n \qquad 7.13$$

According to the reliability requirement of the estimate of $F$, we can chose fe to be 2*sem($F$) giving about a 95% probability that $F$ will lie between $s/n + fe$ and $s/n - fe$, or fe to be 3*sem($F$) which increases the chance to 99%. (Confidence limits).

Solving for these two cases, we get:

$$n = 4*P*(1 - P)/fe\text{^}2 \qquad (fe = 2*sem(F)) \qquad 7.14$$

or

$$n = 9*P*(1 - P)/fe\text{^}2 \qquad (fe = 3*sem(F)) \qquad 7.15$$

If the expected value of $P$ is known approximately, this can be substituted in the relevant equation and $n$ determined. However, if an approximate value of $P$ is not available, both 7.14 and 7.15 have a maximum for $P = 0.5$, giving maximum values for n of $1/fe\text{^}2$ and $9/4/fe\text{^}2$ respectively. Suppose $fe$ is 1%, then, in the worst case, for 99% confidence, at least 22 500 particles must be counted.

The above argument is not valid for $P$ close to zero, (rare events) because, as stated, the approximation does not hold and we must use the Poisson distribution function. In this case, because of the characteristics of the distribution, confidence limits are difficult to define, but if, in $n$ trials, a number $s$ of successes has occurred, it is possible to make some statements about the probability of success. Sometimes, when the event is dangerous or unpleasant, it may be necessary to estimate its probability from one isolated occurrence.

A Poisson distribution is described by the following relations, where the probability of s successes out of n trials is:

$$P(n,s) = mn\text{^}s*\exp(-mn)/s! \qquad 7.16$$

the mean $mn = n*P$ and variance $vr = mn$ 7.17

where $P$ is the probability of the event.

The only estimate of mn available is s and to derive an improved value, the following method may be used, providing that it is possible to make further trials.

Suppose that in n trials s successes are found and that an extra m trials produce another success, then from 7.16, putting $P(n,s)$ equal to $1/n$

$$1/n = mn^s*\exp(-mn)/s! \qquad 7.18$$

and

$$1/(m + n) = mn*(s + 1)*\exp(-mn)/(s + 1)! \qquad 7.19$$

giving

$$mn \approx s*n/(m + n) \qquad 7.20$$

or

$$P \approx s/(m + n) \qquad 7.21$$

Equation 7.21 may be regarded as giving a lower limit for $P$ and the ratio $s/n$ as giving an upper limit. These approximations are very rough, but, with so little information, there is not much else to be done.

The definition of a rare event depends on how the event is regarded. For example, although earthquakes above a certain level of intensity are not common on a daily basis in Japan, they cannot be said to be unexpected when the time interval is ten years. Similarly, diamonds of about 1 mm diameter occur at a concentration of approximately 1 per 1E8 particles, or about 1 diamond in 300 kg. When, however, a plant has a daily throughput of say, 5000 Tonne, then more than 15000 stones will be found each day.

# 7.03 Measures of central tendency

There is a significant difference between the expressions "typical" and "average", but they are quite often used interchangeably in imprecise discussion. The typical human being has two legs, two eyes and two arms, but the average person is less well endowed. To be sure, the average number of legs is not very far from two, but in a world population of 2E9, the difference is important. This is another case where an approximation can be used adequately in some applications, but will fail miserably in another.

In experimental work the objective is frequently to establish a characteristic value of a quantity which can then be used in other applications. The general method is to carry out a number of trials and then to determine a central value $x$ from these measurements. This is then regarded as representative of the quantity being sought. There are several possibilities and the particular one chosen should depend on the application, the required precision, the type of data available and the measurement system.

As mentioned above, it was tacitly assumed earlier, that most experimentally derived values were normally distributed about a mean. When the data were examined, it was often found that there were an uncomfortable number of "outliers" unexpectedly far from the mean. One technique was then to regard these as due to mistakes or malfunction and discard them, subsequently recalculating the mean from the remaining data. Later this practice was frowned upon as being unjustified and the total mean was reinstated. It is, however, good modern practice to examine the data and the objective to find out which method is best suited to the problem under consideration.

### 7.03.1 Market characteristics for a given product

For a product designed for a specific market, the important characteristics are not those of the average buyer, but those which occur most frequently in the consumer, because these are most likely to influence the sales. In a market survey for the product, the indicated central value is the mode.

The mean has several advantages as a representative quantity. It is unbiassed, is easily calculated, and the product of the sample mean and n is close to the true value of $\Sigma(x)$. For systems using part of a population, however, the median is often better because it is not so strongly affected by valid events lying far from the mean. Thus, in economic surveys, the median income of a given group is usually chosen for comparison purposes. The average income is larger because of the presence of a few very well paid people, while the most common or modal income is likely to be at the lower end of the scale.

## 7.04 Presentation of Data

Owing the increasing use of automatic methods of data collection, the number of events that must be recorded and analysed in a single experiment has grown drastically, and there are some useful approximations that can be applied which are not valid for smaller numbers.

### 7.04.1 Histograms

For an Histogram presentation, the observed variable is divided into sections (preferably of equal size) and the number of observations in each section recorded. The resulting numbers are then plotted in the form shown in Figure 7.01. The following approximations are used:

a). All the observations in a given section are treated as though they were concentrated at the mid point of the interval.

b). The ratio of the number of observations in an interval to the total number, the relative frequency (*fr*), is taken to be close to the probability of the occurrence of an event in the interval.

c). If the data come from a continuous distribution, the ordinate of the density curve at the mid point of the interval is approximately equal to *fr* divided by the interval width.

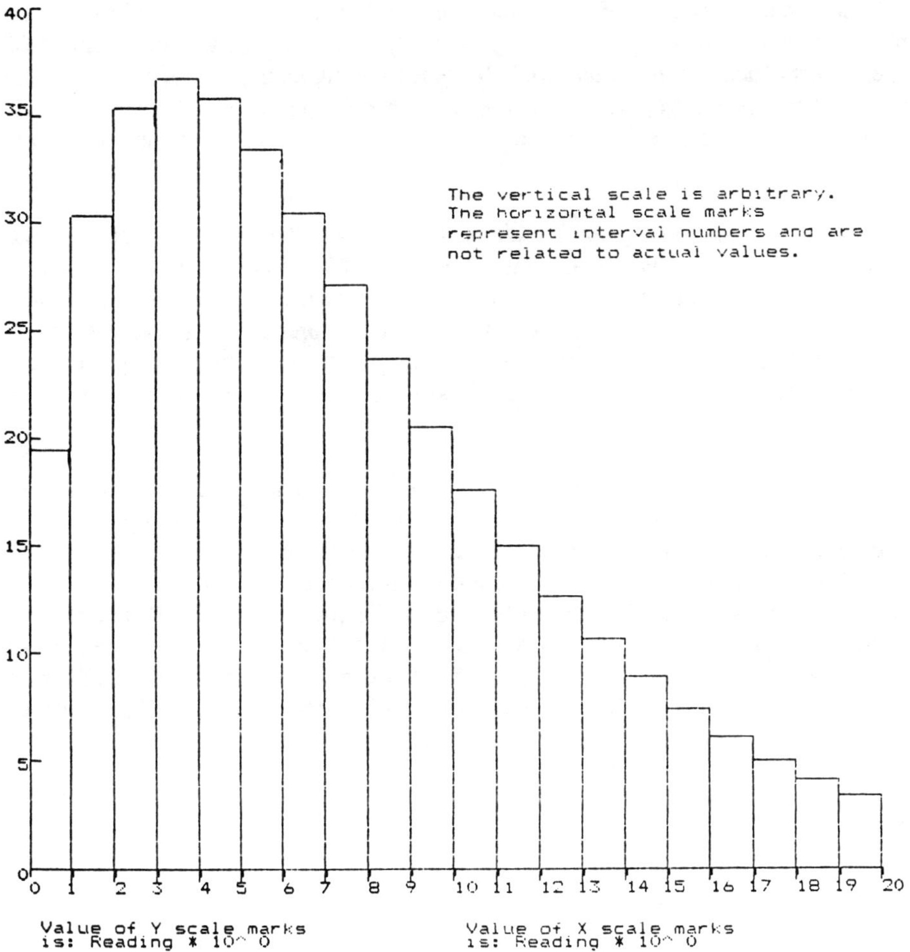


**Fig. 7.01** Histogram Example (Chisquare Distribution)

These approximations are called estimates by statisticians to distinguish them from the mathematical approximations that are necessary to deal with the processing of the values.

As the number of events becomes greater, we expect the estimates to approach limiting values. The errors due to mathematical manipulation may also change, but the two are independent.

To obtain a satisfactory compromise between the precision of the ordinate and the width of the interval, the number of observations in any interval should not be less than 5, unless it is known that the associated probability is so low that the total number of experiments to achieve this value would become prohibitive. In general, even when the number of events is large, there is little point in choosing more than 20 intervals as the overall precision increases approximately as the square root of the number of events recorded.

If the smallest probability in any section is, say, 0.01 and the largest 0.4, then with 20 sections and a simple form of density function, about 2000 observations are necessary. If there are long tails to the function which may be important, then the number may increase to more than 10 000 to achieve satisfactory precision, but increasing the number of intervals for the same number of counts does not help except near the maximum. It was for this reason that 19 ordinates were chosen for the POLYFIT routine.

When we are dealing with amounts of data of this magnitude, and the parent distribution is either continuous or may effectively be considered so, then it pays to fit a curve to the histogram in order to estimate the various characteristics of the distribution. As most density functions are transcendentals, it is necessary to use a relatively high order polynomial to obtain a satisfactory fit, but the 10 coefficients calculated from POLYFIT are usually sufficient. Fig 7.02 shows a chisquare density distribution plotted from the equation, compared with a curve fitted by POLYFIT from an histogram derived from 4211 computer generated points. Table 7.01 lists the values for several statistics obtained from both data sets.

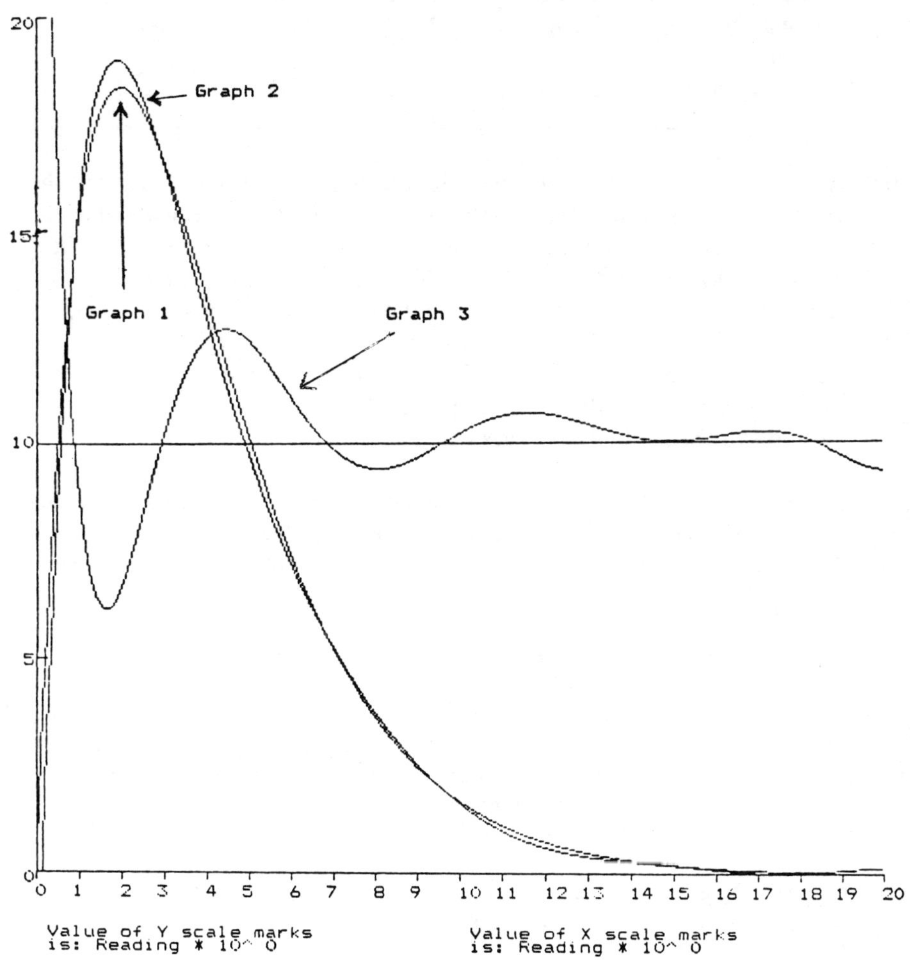


**Fig. 7.02** Comparison between curve fitted to simulation and true curve

**Table 7.01** This table shows the comparison between characteristics derived from raw data and those obtained from a curve fitted to the corresponding histogram. The data were obtained from a computer simulation of a chisquare distribution for 4 degrees of freedom.

| *Statistic* | *Value from:* *Raw Data* | *Fitted curve* | *% Difference* |
|---|---|---|---|
| Mean | 3.918 | 3.963 | 1.15 |
| Median | 3.296 | 3.261 | 1.06 |
| Mode | – | 1.919 | – |
| Std. Deviation | 2.790 | 2.882 | 3.3 |

At first sight it appears as though the fitted curve is not very helpful. However, the standard error of the mean of the data of Table 7.01 is 1.1 % of the mean and thus the 95% confidence limits for the mean lie at about 2% either side of the observed value. This shows that the fitted curve is not unreasonable, but, in any case, the mean, median and standard deviation would be estimated from the raw data. The mode is in a quite different category, and it is difficult to determine a satisfactory value directly. Differentiating the polynomial found by POLYFIT and equating to zero gives a very close estimate of the mode when the resulting equation is solved. The routine FALSPOS can readily be used for this purpose.

The procedure outlined above requires an histogram as a first step and the following short routine finds the numbers in each category, determines the estimated probability density for the mid-interval points, calculates the mean x and puts the values into arrays suitable for POLYFIT.

```
SUBROUTINE HIST.PSD(m,n,x(n),xbar,h(m),x(m),xmax,xmin,ivl)

comment: the array x(n) contains the n+1 data entries from
which the histogram is to be formed. The extreme values,
xmax and xmin are found and xbar determined from their
mean. The range is then divided into an odd number of
intervals m+1 and the data allocated to nh(m). The number
in each interval is then divided by the product of the
interval width and the total number n, and stored in the
array h(). The mid-point of the interval is stored as xx().
endcomment

CONST:   INT m,n;              REAL x(n),lo,hi;
VAR:     INT i,j,k,nh(m);      REAL ivl,xx(m),xmin,xbar,xmax;
                               REAL h(m),hlt,dlt,s;
         BEGIN
```

```
comment: lo and hi are, respectively, the largest negative
and positve values obtainable in the machine. They form
the initial values for the calculation of xmin and xmax.
endcomment

         xmax = lo; xmin = hi;
            DO (i = 0; i = n; i = i+1;)
               IF (x(i)>xmax)
                 xmax = x(i);
               ELSE
                 IF (x(i)< xmin)
                   xmin = x(i);
               ENDIF
            CONTINUE
         xbar = (xmax+xmin)/2;
         ivl = (xmax-xmin)/(m+1);
         dlt = xmin;
         hlt = xmin+ivl;
            DO (i = 0; i = m; i = i+1;)
              xx(i) = (dlt+hlt)/2;
               DO (j = 0; j = n; j = j+1;)
                 IF (x(j)>dlt & x(j)<=hlt)
                   nh(i) = nh(i)+1;
                 ENDIF
               CONTINUE
              dlt = hlt;
              hlt = hlt+ivl;
            CONTINUE
             s = 1/ivl/(n+1)
               DO (j = 0; j = m; j = j+1;)
                h(j) = nh(j)*s;
               CONTINUE
         RETURN (h(),nh(),xx(),xbar,xmax,xmin,ivl)
       END
```

Apart from the histogram, data can also be presented in the form of cumulative frequency or probability curves and this is often done for size distributions. In order to accommodate a large range of sizes it has become general practice to use a logarithmic scale for the abscissa. Unfortunately, many experimenters tend to use a logarithmic scale for the probability axis as well and this is a prolific source of inaccuracy. As discussed in earlier chapters, logarithmic presentation should be used with extreme caution. If a logarithmic x axis is required, the individual values in the x() array should be converted first before any manipulation is performed. The resulting range may then be treated as an ordinary variable.

In a similar manner to that for the histogram, a curve can be fitted to the sums of the estimated probability densities from HIST and this will be an approximation to the distribution function. For unknown distributions, this method gives a useful technique for estimating intermediate probability levels, but the precision will depend strongly on the amount of data available. As discussed by Birnbaum (Bib 14), Kolmogorov developed a method for finding the maximum difference between the actual distribution function and the observed cumulative distribution. To produce a difference not greater than 1% with a confidence level of 95%, almost 20 000 data points are required.

Thus, although, as we have seen, the curve fitting method has its limitations, the errors so introduced are not significantly greater than those due to the probabalistic nature of the data themselves. Fig 7.03 shows a curve fitted to simulated cumulative Normal data compared with the original values.

It is better to derive the mode from the histogram curve instead of attempting to differentiate the cumulative function twice, because of the inevitable loss of accuracy, although the cumulative curve seems to fit the data more closely.

# 7.05 Approximations for Central Values

## 7.05.1 Methods for the mean

For most non-symmetrical distributions with one mode, the median lies between the mode and the mean. Procedures for estimating each of these quantities are given in many statistical texts, but there are some approximation methods which may not be so well known.

## 7.05.2 What is the effect if an extreme value is neglected ?

Suppose that a sample mean (smn) has been found for n samples and the sample standard deviation is ssd. On examining the data, one value xe, lies more than 3*ssd from the mean. Now, if the data are Normally distributed this is unlikely, but, if n is not very large, say less than 20, it may not be possible to decide with sufficient confidence what the actual distribution is. If xe is discarded, the new mean mn1 is given by:

$$mn1 = (n*smn - xe)/(n - 1) \qquad 7.22$$

and the approximate change by:

$$\text{change} = (smn - xe)/n \qquad 7.23$$

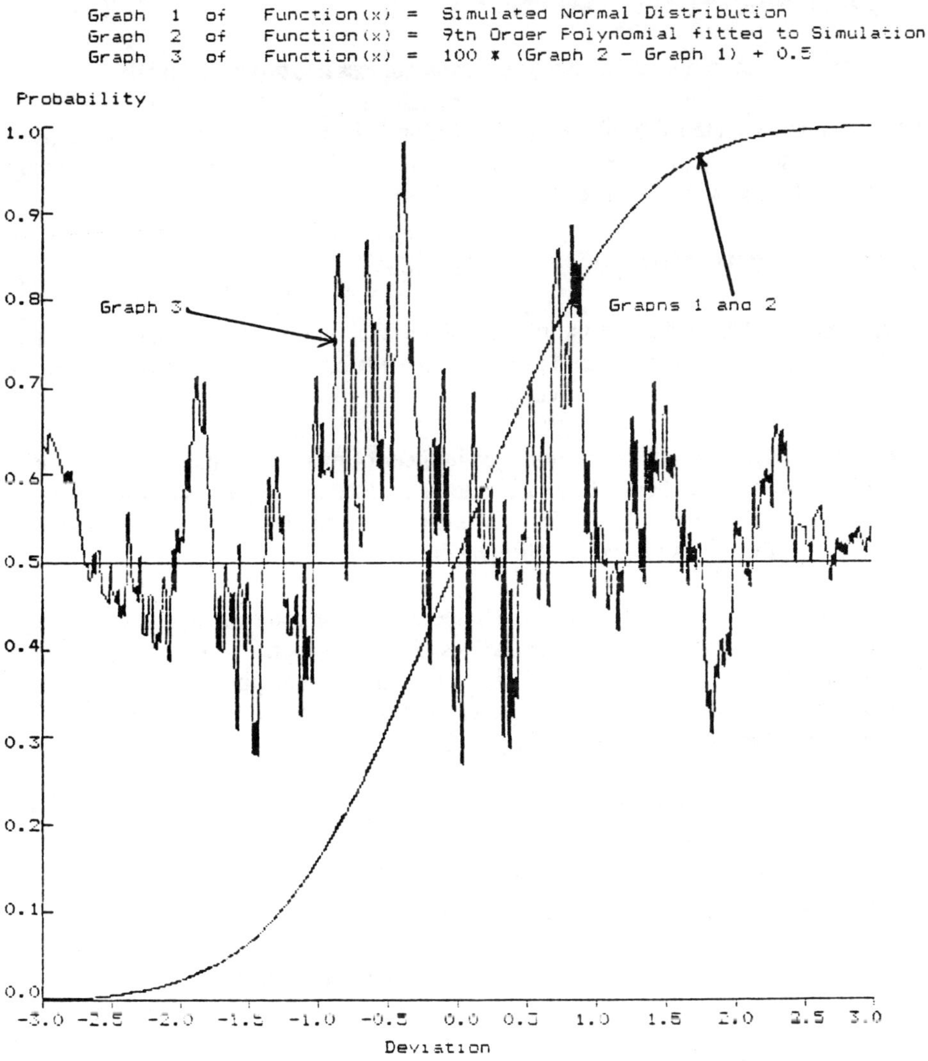

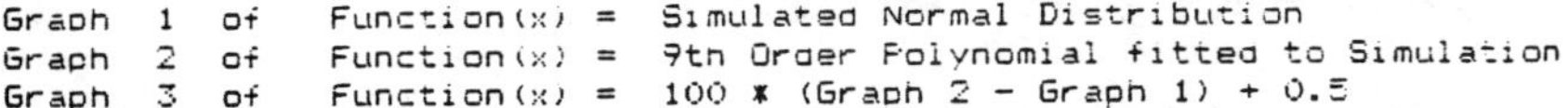


Note: as the curves of graphs 1 & 2 are so close on the scale, the percentage difference is plotted as graph 3, centred on the 0.5 probability value.

Formula 7.32 in exercises 7. No. 3 is used for the simulation of the deviates

**Fig. 7.03** Simulated normal Distribution and fitted Polynomial

The change in the standard deviation is roughly:

$$\text{sd change} \approx xe*(2*smn - xe)/(2*ssd*n) \qquad 7.24$$

These relations should be used only as indicators unless n is greater than 50.

For very approximate work, almost guesstimation, a first estimate of the mean is given by the median or by the mid point of the range and if the distribution is reasonably symmetrical these will be quite close.

## 7.05.3 Evaluation of the median

It was stated in Chapter 1 that the median of an even number of data points does not exist and, although this statement is strictly true, it is convenient to define a fictitious number which can serve the same purpose.

Suppose we have $2*n$ data points arranged in order of magnitude, then, in the absence of further information, the median value can be defined as the value corresponding to half way between the $n$th and the $(n+1)$th point. If the number of observations is odd (2*n+1), then the median is the nth observation.

In the case of a continuous distribution, the median is that point which divides the area under the density curve into two equal halves. This corresponds to the value at which the probability of occurrence of all the values up to that point is one half. It is sometimes convenient to estimate the median from the fitted cumulative curve by taking the x value of the 50% probability level.

## 7.05.4 Determination of the mode

If a fitted curve is not available, but an histogram has been made, the mode may be approximately estimated by the following method:

Let the kth section contain the mode, then, because the form of the curve through the mode must be nearly a parabola, we have:

$$xm = ivl*(h(k-1) - h(k+1)/(h(k) - h(k-1))/4 \qquad 7.25$$

where $xm$ is the distance of the mode from the centre of the kth interval and the other symbols have the same meanings as in section 7.04

There is no simple method for finding a confidence limit for this value, but, if the number of events is of the order of 500 or more, then the error of the mode determined in this way will probably be less than $ivl/4$.

# 7.06 Approximations using Order Statistics

When the distribution of the available data is clearly not Normal, it is helpful to be able to put limits on the determined values of the derived statistics and the following approximations may be found useful:

Let the n samples be arranged in order of magnitude and denote the *k*th sample by U(k) and the probability of finding that sample up to the point xk be Pu. Remarkably, the density function of *Pu* as a function of *Pu* turns out to be independent of the distribution and has the form of a Beta function. Now if *n* is large, say of the order of 100 or more, the Beta function can be approximated by a Normal distribution with a mean *z* given by:

$$z = (Pu - Pum)/Pum \qquad 7.26$$

and a variance vz

$$vz = (1 - Pum)/(n*Pum) \qquad 7.27$$

where *Pum* is the modal value of the Beta function.

If a density curve has been fitted to the data, *Pum* can be taken as the area under the curve up to *xk*. From this we can say that, with 95% confidence the sample *U*(*k*) has an x value given by:

$$xk\text{-}(2*\sqrt{}(A*(1 - A)/n)/y) < xk < xk + (2*\sqrt{}(A*(1 - A)/n)/y) \qquad 7.28$$

where *A* is the area under the curve and *y* is the ordinate at the point xk.

The limits for the mode given by this relation, calculated for the chisquare simulation described in section 7.03 are:

$$1.82 < \text{mode} < 2.04 \qquad 7.29$$

The mode determined from the equation of the distribution is 2.

Similarly, for the histogram, if *s* is the number of entries in the *j*th section and *Ps* is the probability of an event falling in that section then the expectation (mean) of *Ps* is given by:

$$\text{Expectation}(Ps) = s/(n + 1) \qquad 7.30$$

and

$$\text{variance}\ (Ps) = s*(n - s + 1/((n + 1)^2*(n + 2)) \qquad 7.31$$

As $n$ and $s$ become very large equations 7.30 and 7.31 tend to the binomial expressions:

As s becomes very small at the tails of the distribution the variance of the ratio $Ps/(s/n)$ tends to $1/s$ and the reliability of any assessment technique at these points is very low. There is only one way of improving evaluation for such sections and that is to perform additional experiments or trials. If there is some information about the form of the function, this can, of course be used, but decisions based on probabilities derived from the tail of a distribution should be regarded as suspect, unless there is adequate supporting evidence.

# Exercises 7

1. Using the random generator of your computer, construct a random variable from the mean of the sum of a number of consecutive random numbers.

   a). Vary the number from 5 to 30 and determine the mean and standard deviation for several runs using different total numbers of trials.

   b). Compare the calculated standard deviations with that for a Normal distribution with the same mean.

   c). Run the program once more with a different seed for the random number starting value and find whether the observed value is within the limits predicted by the 95% confidence level.

2. A commuter travels every day 20 km on a road with 22 sets of traffic lights. The probabilty that a given set will be green in his/her favour is 0.25 and the traffic moves with a speed of 40 km/hr between the lights. If the light is red, the waiting time is about 30 seconds.

   a). Is the Normal distribution a satisfactory approximation for the binomial in this case ?

   b). What is the average speed for the journey ?

   c). What is the probability that the journey will not have a single red light ?

   d). What is the probability that the journey will not have a single greeen light ?

e). What are the limiting times for the journey in cases c and d and the average time?

f). What difference would it make to the average time of the journey if the speed between the lights were increased to 60 km/hr ?

3. It can be shown that, if *RD1* and *RD2* are independent, uniformly distributed random numbers between 0 and 1, then *y* given by

$$y = \cos(2*\pi*RD2) * \sqrt{(-2*\log(RD1))} \qquad 7.32$$

is a random number with a standard Normal distribution.

a). Write a program using successive values of the random number generated by the computer to produce several thousand such pseudo-Normally distributed numbers and store them on disk.

b). Estimate the probability of finding a number between 0 and 1 from these data and compare the result with that given by the usual tables.

c). How large does the sum of exercise 1 have to be to give as good a result as that from equation 7.32 ?

4. In a shop, analysis of the books over a relatively long time shows that the number of customers per day is approximately Normally distributed with a mean of 500 and a standard deviation of 100. The amount spent per customer, however, is more nearly a chisquare distribution with 4 degrees of freedom (mean = £4, mode = £2).

A chisquare distribution with 4 degrees of freedom can be constructed by taking the squares of four successive values of a normal random number and adding them to produce a new random variable.

a). Construct the density distribution of the random variable - total amount taken per day - by multiplying values from each of the distributions together for each day, finding the sum and storing the result. Use the methods outlined in Chapter 7 for fitting an appropriate curve.

b). Check that the mean of this distribution is close to £2000.

c). Determine the most probable daily turnover.

d). Integrate the density function and hence determine the chance of taking more than £10 000 in one day ?

e). Shortly before the end of the financial year, an estimate is required for the year-end balance. Which central measure should be used to predict the turnover for the remaining time ?

# BIBLIOGRAPHY

Where the publication cited has particular importance for the subject or has noteworthy relevance a comment is given.


1. Press W.H. et al 1986 *Numerical Recipes* (Cambridge UP)
comment: this book is outstanding and is almost essential for serious workers with a good programming background. It is rather advanced and needs thorough study to make the best use of it. endcomment.

2. Kernighan B.W. 1978 *The elements of Programming Style* (McGraw-Hill)

3. Ralston A. and Rabinowitz P. 1978 *A First Course in Numerical Analysis* (McGraw-Hill)

4. Wolfe M.A. 1972 *A First Course in Numerical Analysis* (Van Nostrand Reinhold)
comment: a very useful introduction to numerical analysis, but with a terse style. Useful examples are given. endcomment.

5. Fairbrother H. 1964 *Mathematical Notes for Engineering Students* (Private communication)

6. Lawson C.L. and Hanson R.J. 1974 *Solving Least Squares Problems* (Prentice-Hall)

7. Luke Yudel L. 1975 *Mathematical Functions and their Approximations* (Academic Press)

8. Chatfield C. 1978 *Statistics for Technology* (Chapman and Hall)
comment: This a very readable book for the engineer which contains many examples and techniques. It does not particularly stress the use of computing techniques. endcomment

9. Hart John F. et al 1968 *Computer Approximations* (Wiley)

10. Hastings C. 1955 *Approximations for Digital Computers* (Princeton University Press)

11. Acton F.S. 1970 *Numerical Methods that work* (Harper and Row)

12 Westlake J.R. 1968 *A Handbook of Numerical Matrix Inversion etc.* (Wiley)

13. Golub G.R. and Van Loan C.F. 1983 *Matrix Computations* (John Hopkins University Press)

14 . Birnbaum Z.W. 1963 Probability and Mathematical Statistics (Harper and Row) comment: A rather advanced text for the mathematically inclined, but it has some useful methods and techniques not generally included in the more elementary works. endcomment.

15. Noether G.E. 1978 *Introduction to Statistics* (Houghton Mifflin) comment: this book has been included in the bibliography because it adopts a non-parametric approach which is receiving more attention due to the difficulty of determining distribution functions in cases where the data are sparse. endcomment.

# Appendix A

# INDEX OF PSEUDO-CODE PROGRAMS AND SUBROUTINES

The entries are listed in alphabetical order as given in the contents table below, but the page number refers to the page in the text where the code can be found.

With the exception of the suffix .PSD, the names of the programs are the same as for the BASIC listings. The program ARRYCALC.BAS does not have a Pseudo-code equivalent as the character handling procedures differ so much in the various languages.

# Appendix B

# BASIC LISTINGS

## B 1. INDEX OF BASIC PROGRAMS AND SUBROUTINES

The entries are listed in alphabetical order as given in the contents table below.

With the exception of the suffix .BAS, the names of the programs are the same as for the Pseudo-code algorithms. The program ARRYCALC.BAS does not have a Pseudo-code equivalent because the character handling procedures differ too much in the various languages.

The comments given for the Pseudo-code programs apply also to the BASIC versions and are not repeated in the listings, except where considered necessary.

## B 2. LISTING OF BASIC PROGRAMS AND SUBROUTINES

ACCSER.BAS

This short routine is used for Fairbrother's accelerated series summing technique. K is the number of terms in the original series.

```
10      REM ACCSER.BAS FOR SERIES CALCULATION
20      DEFINT I-N  :K = 5     : REM SET K
30      DIM A(K),SUM(K)
40      FOR I = 0 TO K
50      READ A(I)
60      NEXT I
65      S = A(0)
70      FOR I = 1 TO K
75      R = A(I-1)/(A(I-1)-A(I))
80      SUM(I) = S+R*A(I)
90      S = S+A(I)
100     NEXT I
110     FSUM = SUM(K-1)+(SUM(K)-SUM(K-1))*R
120     DATA (Enter K+1 numerical terms here)
130     PRINT FSUM
140     END
```

ARRYCALC.BAS

These subroutines use character arrays as storage. The numbers are in ASCII form and manipulated using the string instructions common to most BASIC dialects. The length the numbers is limited only by the maximum length of a string in the dialect being used

# General notes for the subroutines for ARRYCALC

All the input numbers must be integers; the division routine can produce an output with decimal fractions. The addition of the decimal point is possible for all routines, but it adds to the complexity and reduces the speed. For decimal fractions it is usually simpler to remove the point and adjust the length of the numbers accordingly.

The following 5 lines are required for all the routines and should appear in a main program before the routines are called.

```
10   DEFDBL P
20   OPTION BASE 1
30   DEFINT H-N:MN = 18:KTOT = -10
40   R$="0":MK = INT((MN+2.9)/3):MT = MK*3
50   DIM NUM1(MN),NUM2(MN),F1$(MN),F2$(MN),LP(256)
```

### a) Division Subroutine

The two numbers must be integers in the form of ASCII strings; the numerator is A1$ and the denominator is A2$. They can be of any length up to the allowable string maximum (usually 256 characters). The input values are converted into the interim variables D1$ and D2$ for the process. The output is DVD$ and the remainder RD$: A minimum decimal exponent for the remainder is given as KTOT which is set to -10 in the listing. The routine is ordinary long division essentially and uses the subtraction routine. Multiple subtraction is easier to program than using the multiplication method with trial and error and is probably faster.

```
1000  DIV:REM DIVISION SUBROUTINE
1060  MV = 0
1080  D1$ = A1$:D2$ = A2$:LD = LEN(A1$)
1090  WHILE LEN(D1$)<LEN(D2$)
1100  D1$ = D1$+R$:MV = MV-1:WEND
1110  B$ = D1$:C$ = D2$
1120  KO = MV+1:DVD$="":LE = LEN(C$)
1130  A3$ = LEFT$(B$,LE):B$ = A3$:RD$ = B$
1140  WHILE (VAL(RD$)<>0 OR LE<=LD) AND KO>KTOT
1150  NV = 0:FR$ = ""
1160  IF LEFT$(B$,1) = "0" THEN G$ = MID$(B$,2):B$ = G$
1170  WHILE FR$<>"-":G1$ = B$:G2$ = C$
1180  GOSUB SUB:NV = NV+1:B$ = OPT$:WEND
1190  B$ = G1$
1200  NV = NV-1:DVD$ = DVD$+RIGHT$(STR$(NV),1)
1210  WHILE LEFT$(DVD$,1) = "0":Q$ = MID$(DVD$,2)
```

```
1220  DVD$ = Q$:WEND
1230  LE = LE+1:IF LE = LD+1 THEN DVD$ = DVD$+"."
1240  WHILE LE<=LD:A$ = B$+MID$(D1$,LE,1):B$ = A$
1250  RD$ = B$:GOTO 1270:WEND
1260  RD$ = B$:B$ = B$+R$:IF LE>LD THEN KO = KO-1
1270  IF VAL(RD$) = 0 AND LE>LD THEN 1290
1280  WEND
1290  IF VAL(STR$(KO)) = 0 THEN 1310
1300  RD$ = RD$+" *10^"+STR$(KO)
1310  LPRINT A1$;" / ";A2$;" = ";DVD$;" :REMAINDER = ";RD$
1320  RETURN
```

### b) Multiplication subroutine

This routine is for integers only.

For multiplication the two input strings are G1$ and G2$. The working dummy variables are SNUM1$ and SNUM2$. Because of the speed of the normal multiplication system, it was found to be better to separate the numbers into groups of three, (lines 1400 to 1460), multiply (lines 1470 to 1580), and then add the separate products with the appropriate shifts.

The integer arrays NUM1() and NUM2() are used for holding the interim conversions.

The output product is OPT$

```
1330  REM MULTIPLICATION ROUTINE
1340  MUL:
1350  SNUM1$ = G1$:SNUM2$ = G2$
1360  IF LEN(SNUM2$)>LEN(SNUM1$) THEN SWAP SNUM1$,SNUM2$
1370  LE1 = LEN(SNUM1$):LE2 = LEN(SNUM2$)
1380  MT = INT((LE1+2.9)/3):MU = INT((LE2+2.9)/3)
1390  MMT = 1+INT((LE1+LE2+2.9)/3)
1400  X$ = SNUM1$:X2$ = SNUM2$:I = 1:J = 1
1410  WHILE LEN(X$)>0:NUM1(I) = VAL(RIGHT$(X$,3))
1420  Y$ = LEFT$(X$,LEN(X$)-3)
1430  X$ = Y$:I = I+1:WEND
1440  WHILE LEN(X2$)>0:NUM2(J) = VAL(RIGHT$(X2$,3))
1450  Y$ = LEFT$(X2$,LEN(X2$)-3)
1460  X2$ = Y$:J = J+1:WEND:
1470  NK = 1000:LO = 0:KAR = 0
1480  FOR I = 1 TO MMT:LP(I) = 0:NEXT I
1490  FOR K = 1 TO MT
```

```
1500  FOR KK = 1 TO MU
1510  P = NUM1(K)*NUM2(KK)
1520  HI = INT(P/NK):LO = P-HI*NK:LP(KK+K-1) = LP
      (KK+K-1)+LO
1530  IF LP(KK+K-1)<NK THEN 1560
1540  KR = INT(LP(KK+K-1)/NK):LP(KK+K-1) = LP
      (KK+K-1)-NK*KR
1550  HI = HI+KR
1560  LP(KK+K) = LP(KK+K)+HI
1570  IF LP(KK+K)<NK THEN 1600
1580  KAR = INT(LP(KK+K)/NK):LP(KK+K) = LP(KK+K)-NK*KAR
1590  LP(KK+K+1) = LP(KK+K+1)+KAR
1600  NEXT KK:NEXT K:OPT$ = ""
1610  FOR J = MMT TO 1 STEP -1:TR$ = RIGHT$(STR$(LP
      (J)+NK),3)
1620  OPT$ = OPT$+TR$
1630  NEXT J
1640  WHILE LEFT$(OPT$,1) = "0":PT$ = MID$(OPT$,2)
1650  OPT$ = PT$  :WEND
1660  RETURN
```

## c) Addition subroutine

This routine is for integers only.

Again the two input strings are G1$ and G2$ and the working dummy variables are SNUM1$ and SNUM2$. The triplet separation is used, (lines 1710 to 1790), and the numbers are added in lines 1800 to 1890.

The integer arrays NUM1() and NUM2() are used for holding the interim conversions and the output is OPT$.

```
1670  REM ADDITION SUBROUTINE
1680  ADD:SNUM1$ = G1$:SNUM2$ = G2$
1690  IF LEN(SNUM1$)<LEN(SNUM2$) THEN SWAP SNUM1$,SNUM2$
1700  LE1 = LEN(SNUM1$):LE2 = LEN(SNUM2$)
1710  MT = INT((LE1+2.9)/3):MU = INT((LE2+2.9)/3)
1720  MMK = 1+MT:X$ = SNUM1$:X2$ = SNUM2$:I = 1:J = 1:NK =
      1000
1730  FOR JI = 1 TO MT:NUM1(JI) =0:NUM2(JI) =0:LP(JI)
      =0:NEXT JI
1740  WHILE LEN(X$)>0:NUM1(I) = VAL(RIGHT$(X$,3))
1750  Y$ = LEFT$(X$,LEN(X$)-3)
```

```
1760  X$ = Y$:I = I+1:WEND
1770  WHILE LEN(X2$)>0:NUM2(J) = VAL(RIGHT$(X2$,3))
1780  Y$ = LEFT$(X2$,LEN(X2$)-3)
1790  X2$ = Y$:J = J+1:WEND:
1800  FOR I = 1 TO MMK: LP(I) = 0:NEXT I
1810  FOR K = 1 TO MT
1820  P = NUM1(K)+NUM2(K)
1830  HI = INT(P/NK):LO = P-HI*NK:LP(K) = LP(K)+LO
1840  LP(K+1) = LP(K+1)+HI
1850  NEXT K:OPT$ = ""
1860  FOR J = MT TO 1 STEP -1:TR$ = RIGHT$(STR$(LP
      (J)+NK),3)
1870  OPT$ = OPT$+TR$
1880  NEXT J:IF LP(MMK) = 1 THEN OPT$ = "1"+OPT$
1890  WHILE LEFT$(OPT$,1) = "0"
1900  PT$ = MID$(OPT$,2):OPT$ = PT$  :WEND
1910  RETURN
```

### d) Subtraction subroutine

This routine is for integers only.

As before the two input strings are G1$ and G2$ and the working dummy variables are SNUM1$ and SNUM2$. The triplet separation is used, (lines 1710 to 1790), and addition with nines complement is employed. The numbers are added in lines 2170 to 2220. Leading zeros are removed in lines 2230 to 2270.

The integer arrays NUM1() and NUM2() are used for holding the interim conversions and the output is OPT$.

```
1920  REM SUBTRACTION SUBROUTINE
1930  SUB:NC = 999:SNUM1$ = G1$:SNUM2$ = G2$
1940  LE1 = LEN(SNUM1$):LE2 = LEN(SNUM2$)
1950  FR$ = ""
1960  IF LE1>LE2 THEN 2050
1970  IF LE1<LE2 THEN 2040 :XL$ = SNUM1$:YL$ = SNUM2$
1980  XXL = VAL(LEFT$(XL$,4))
1990  YYL = VAL(LEFT$(YL$,4))
2000  WHILE XXL<=YYL AND LEN(XL$)>0
2010  IF VAL(LEFT$(XL$,4))<VAL(LEFT$(YL$,4)) THEN 2040
2020  XY$=MID$(XL$,5):YY$ = MID$(YL$,5):XL$ = XY$:YL$ =
      YY$
2030  WEND:GOTO 2050
```

```
2040  SWAP SNUM1$,SNUM2$:FR$ = "-"
2050  MK = INT((LEN(SNUM1$)+2.9)/3)
2060  X$ = SNUM1$:X2$ = SNUM2$:I = 1:J = 1
2070  FOR JI = 1 TO MK:NUM1(JI) = 0:NUM2(JI) = 0:NEXT JI
2080  WHILE LEN(X$)>0:NUM1(I) = VAL(RIGHT$(X$,3))
2090  Y$ = LEFT$(X$,LEN(X$)-3)
2100  X$ = Y$:I = I+1:WEND
2110  WHILE LEN(X2$)>0:NUM2(J) = VAL(RIGHT$(X2$,3))
2120  Y$ = LEFT$(X2$,LEN(X2$)-3)
2130  X2$ = Y$:J = J+1:WEND:
2140  FOR I = 1 TO MK:NUM2(I) = NC-NUM2(I)
2150  NEXT I
2160  NK = 1000:MMK = MK+1
2170  FOR I = 2 TO MMK:LP(I) = 0:NEXT I:LP(1) = 1
2180  FOR K = 1 TO MK
2190  LP(K) = NUM1(K)+NUM2(K)+LP(K)
2200  IF K = MK THEN 2220:HI = INT(LP(K)/NK)
2210  LP(K+1) = LP(K+1)+HI
2220  NEXT K:OPT$ = ""
2230  FOR J = MMK TO 1 STEP -1:TR$ = RIGHT$(STR$(LP
      (J)+NK),3)
2240  OPT$ = OPT$+TR$:NEXT J:W$ = LEFT$(OPT$,1)
2250  WHILE W$ = "0":Q$ = MID$(OPT$,2):OPT$ = Q$:W$ =
      LEFT$ (OPT$,1)
2260  IF LEN(OPT$) = 1 AND W$ = "0" THEN 2270:WEND
2270  OPT$ = FR$+OPT$
2280  RETURN
2290  END
```

BINTERM.BAS

```
10 BINT:
20 I = I-1:  R = R+1
30 IF I> 0 THEN ANS = ANS*X*I/R
40 RETURN
```

## CONVERT.BAS

Converts binary numbers to binary coded decimal.

```
10      ZE% = 48:DIM NUM%(20)
15      PRINT "ENTER NUMBER"
20      INPUT  ANUM : IF ANUM = 0 THEN 70
30      GOSUB CONV
40      FOR K = 1 TO I
50      PRINT CHR$(NUM%(K));
60      NEXT K :PRINT:GOTO 15
70      END
3000    CONV: REM SUBROUTINE FOR BINARY TO BCD CONVERSION
3010    FOR J = 0 TO 20: NUM%(J) = 48 :NEXT J
3020    ANUM1 = ANUM: BNUM = 0
3030    DV = 1: KO = 1: T = 10:T1 = 0.1:ETA = ANUM/1E6
3040    WHILE  ANUM1>=T
3050    ANUM1 = ANUM1*T1 :DV = DV*T
3060    KO = KO+1:WEND
3070    I = 1
3080    WHILE ANUM-BNUM >ETA
3090    IF I = KO+1 THEN NUM%(I) = ASC("."):I = I+1
3100    FM1% = INT(ANUM1)
3110    NUM%(I) = FM1%+ZE% :BNUM = BNUM+DV*FM1%
3120    ANUM1 = ANUM1-FM1%
3130    ANUM1 = ANUM1*T
3140    DV = DV*T1
3150    I = I+1: WEND
3160    I = I-1
3170    IF KO > I THEN I = KO
3180    RETURN
```

## DIVPOL.BAS

```
3000    DIVPOL:
3010    FOR I% = 0 TO N1%
3020    RM(I%) = CF1(I%):CF3(I%) = 0
3030    NEXT I%
3040    FOR J% = N1%-N2%  TO 0 STEP -1
3050    CF3(J%) = RM(N2%+J%)/CF2(N2%)
3060    FOR K% = N2%+I%-1 TO I%  STEP -1
3070    RM(K%) = RM(K%)-CF3(I%)*CF2(K%-J%)
3080    NEXT K%
```

```
3090  NEXT J%
3100  RM(N2%) = 0
3110  RETURN
```

DIVSER.BAS

```
3000  DIVSER:
3010  FOR I% = 0 TO N1%
3020  RM(I%) = COEF1(I%):COEF3(I%) = 0
3030  NEXT I% :K% = 0 :U = COEF2(0)
3032  WHILE COEF2(K%) = 0
3034  U = COEF2(K%+1)
3036  K% = K%+1
3038  WEND
3040  FOR I% = 0 TO N1%
3050  COEF3(I%) = RM(I%)/U
3060  FOR J% = I% TO N1%
3070  RM(J%) = RM(J%)-COEF3(I%)*COEF2(J%-I%)
3080  NEXT J%
3090  NEXT I%
3110  RETURN
```

EXPOM.BAS

```
1000  EXPOM:  REM CALCULATES EXP(X) FROM (1 + X/N)^N
1010  K = 10: REM SET K
1020  Z = 1 + X/N
1030  FOR I = 1 TO K
1040  Z = Z * Z
1050  NEXT I
1060  RETURN
```

****************************************

FALSPOS.BAS

```
10   REM INITIAL VALUES : Insert initial values for ETA
     etc.
20   DEF FNX(G) = (Define equation to be solved).
```

```
1000 FALSPOS:
1010 Y0 = FNX(X0)
1020 Y1 = FNX(X1)
1030 X = X0
1040 X2 = X1
1050 Z = ABS(X2-X)
1060 Z1 = Z
1070 WHILE Z>ETA
1080 IF Z>Z1 THEN 1090 ELSE 1120
1090 PRINT "System does not converge "
1100 PRINT "Check function and start values"
1110 RETURN
1120 Z1 = Z
1130 X2 = X
1140 IF Y0*Y1<0 THEN X0 = X ELSE X0 = X1:X1 = X
1150 X  = X1-(X1-X0)*Y1/(Y1-Y0)
1160 Y1 = FNX(X)
1170 Z = ABS(X2-X)
1180 WEND
1190 RETURN
```

FOURIER.BAS

```
100    REM PROGRAM FOURIER.BAS TO FIT A FOURIER SERIES
120    FSP$ = "(Write function and date here)"
140    DEFINT I-N:  NC = 16  :NH = 10 :NT = NH*NC
150    REM Set necessary constants in this space
160    PI = 3.14159:RNGE = 2*PI
180    DIM A(NH),B(NH)
200    JF = 0:  GOSUB SIMPF
270    A(0) = SUMF/2
280    FOR J = 1 TO NH :JF = 1
300    NC = 16*J
340    H = 2/3/NC: GOSUB SIMPF
360    A(J) = SUMF  :JF = 2
410    GOSUB SIMPF
420    B(J) = SUMF
430    NEXT J
440    LPRINT "   FOURIER COEFFICIENTS FOR ";
460    LPRINT FSP$:LPRINT "        Run Number   ";
       RNUM:LPRINT
480    LPRINT TAB(5);"No.";TAB(15);"A";TAB(35);"B"
500    FOR M = 0 TO NH
```

```
520    LPRINT TAB(3);:LPRINT USING "#####";M;
540    LPRINT USING "+#.###";TAB(12);A(M);:IF M = 0 THEN 580
560    LPRINT USING "+#.###";TAB(32);B(M)
580    NEXT M
600    END
3000   FUNC:
3010   REM Define function here
4000   RETURN
```

HIST.BAS

```
300    REM HISTOGRAM SUBROUTINE : N IS NUMBER OF DATA
       POINTS
310    HIST:
320    LO = 1E-10:HI = 1E10:XMAX = LO:XMIN = HI
330    M = 18
340    FOR I = 0 TO N
380    IF X(I)>XMAX THEN XMAX = X(I)
385    IF X(I)<XMIN THEN XMIN = X(I)
390    NEXT I
400    REM M + 1 IS NUMBER OF INTERVALS :  IVL IS INTERVAL
       WIDTH
410    XBAR = (XMAX+XMIN)/2
420    IVL = (XMAX-XMIN)/(M+1)
430    DLT = XMIN : HLT = DLT+IVL
440    FOR I = 0 TO M :NH(I) = 0
450    XX(I) = (DLT+HLT)/2
460    FOR J = 0 TO N
470    IF X(J)>DLT AND X(J)<=HLT THEN NH(I) = NH(I)+1
480    NEXT J
490    DLT = HLT:HLT = HLT+IVL
500    NEXT I
510    S = 1/IVL/(N+1)
520    FOR K = 0 TO M
530    H(K) = NH(K)*S
540    NEXT K
550    RETURN
```

LOOP.BAS

This program fragment demonstrates the technique for avoiding problems with non-integer indices.

Instead of using a non-integer STEP such as 0.1, for example, the loop should be:

```
10 SUM = 0: T = 0.1: J = 10: K = 1
20 FOR I = 0 TO J
30 SUM = SUM + T*I
40 PRINT I*T, SUM
50 NEXT I
60 END
```

MULPOL.BAS

```
10      MULPOL:
20      R% = 0:S% = 0:T% = 0
30      WHILE T%<(N1%+N2%)
40      FOR R% = 0 TO N1%
50      FOR S% = 0 TO N2%
60      IF T% = R%+S% THEN COEF3(T%) = COEF3(T%)+COEF1
        (R%)*COEF3 (S%)
70      NEXT S%
80      NEXT R%
90      T% = T%+1
100     WEND
110     RETURN
```

PARAFIT.BAS

```
2000 PARA:  REM FITS PARABOLA TO THREE POINTS X(I), Y(I)
2010 I = 3
2020 Z(0) = X(0) - X(1): Z(1) = X(1) - X(2)
2030 Z(2) = X(0) - X(2): Z(3) = X(0) + X(1)
2040 W(0) = Y(0) - Y(1): W(1) = Y(1) - Y(2)
2050 A(2) = (W(0)/Z(0) - W(1)/Z(1))/Z(2)
2060 A(1) = (W(0)/Z(0)) - (A(2)*Z(3))
2070 A(0) = Y(0) - A(1)*X(0) - A(2)*X(0)*X(0)
2080 REM THE COEFFICIENTS A() ARE RETURNED
2090 RETURN
```

POLINT.BAS

```
2000   POLINT: REM SUBROUTINE POLINT (POLAR AREA)
2010   SUMF = RR(0)*(RR(0)+RR(1))+RR(NC)*RR(NC)
2020   FOR I = 1 TO NC-1
2080   SUMF = SUMF+RR(I)*(2*RR(I)+RR(I+1))
2100   NEXT I
2120   SUMF = SUMF*BET/6
2140   RETURN
3000   FUNC:
3020   (Define function in this section)
4000   RETURN
```

POLYFIT.BAS

**Note:** F values are included in this routine as FF (), but they can be omitted if desired. Lines 3270 to 3330 can be used as a separate subroutine - BINOM.BAS for calculating binomial expressions if required.

```
3000   POLYFIT:
3010   YM = 0:YYM = 0 :B(0) = 0:C(0) = 0
3020   XM = X(N/2) : REM N IS ASSUMED TO BE EVEN
3030   FOR L = O TO N
3040   YM = YM+Y(L):YYM = YYM+Y(L)*Y(L)
3050   NEXT L  :A(0) = YM/SQ(0)
3055   FF(0) = 0
3060   D = (X(N)-X(0))/N
3070   YSQ = YYM-YM*YM/(N+1):R(0) = 100
3080   FOR I = 1 TO K
3090   A(I) = 0:B(I) = 0:C(I) = 0
3100   FOR L - 0 TO N
3110   A(I) = A(I)+Y(L)*ZF(I,L)
3120   NEXT L
3130   FF1 = A(I)
3140   A(I) = A(I)/SQ(I)
3160   FF(I) = FF1*A(I)*100/YSQ : R(I) = R(I-1) - FF(I)
3170   NEXT I
3180   TEMD = 1
3190   FOR J = 0 TO K :GOTO 3210
3195   IF FF(J)<0.01*R(0) THEN FF(J) = 0
3200   FF(J) = FF(J) * (N-J) / R(J)
3210   FOR I = J TO K STEP 2
3220   B(J) = B(J)+A(I)*CF(I,J)
```

```
3230  NEXT I
3240  C(J) = B(J)*TEMD:TEMD = TEMD/D
3250  NEXT J
3265  IF XM = 0 THEN RETURN ELSE 3270
3270  FOR I = 0 TO K :BIN(I,0) = 1
3272  FOR J = 1 TO I
3274  BIN(I,J) = BIN(I,(J-1))*(I-J+1)/J
3276  NEXT J :NEXT I
3278  FOR I = 0 TO K :M = K
3280  L=K-I:  TEMX = C(M)*BIN(M,L)
3290  FOR J = L TO 1 STEP -1:M = M-1
3300  TEMX = TEMX*(-XM)+C(M)*BIN(M,(J-1))
3310  NEXT J
3320  C(I) = TEMX
3330  NEXT I
3340  RETURN
4070  END
```

POLYFT6.DAT

```
10    REM DATA FOR POLYFT6
20    DIM A(9),B(9),C(9),R(9),SQ(9),F05(9),BIN(9,9)
30    DIM FF(9),SIG(9),CF(9,9),ZF(9,18),Y(18),X(18)
40    DEFINT I-N
50    YM = 0 :YYM = 0
60    N = 6:K = 6
70    FOR I = 0 TO N
80    READ X(I)
90    NEXT I
100   FOR I = 1 TO K
110   READ F05(I)
120   NEXT I
130   DATA 200,250,300,350,400,450,500
150   DATA 161.45,18.51,10.13,7.71,6.61,5.99 :REM F VALUES
160   FOR J = 0 TO N
170   FOR I = 0 TO K
180   READ ZF(I,J)
190   NEXT I
200   NEXT J
210   DATA 1,-3,5,-1,3,-1,1
220   DATA 1,-2,0,1,-7,4,-6
230   DATA 1,-1,-3,1,1,-5,15
```

```
240    DATA 1,0,-4,0,6,0,-20
250    DATA 1,1,-3,-1,1,5,15
260    DATA 1,2,0,-1,-7,-4,-6
270    DATA 1,3,5,1,3,1,1
400    FOR I = 0 TO K
410    READ SQ(I)
420    NEXT I
430    DATA 7,28,84,6,154,84,924
440    FOR I = 0 TO K
450    FOR J = 0 TO I
460    READ CF(I,J)
470    NEXT J
480    NEXT I
490    DATA 1
500    DATA 0,1
510    DATA -4,0,1
520    DATA 0,-1.166667,0,0.1666667
530    DATA 5.9999975,0,-5.583333,0,0.5833333
540    DATA 0,8.733484,0,-0.6228814,0,0.35
550    DATA -20,0,50.63333,0,-16.91667,0,1.2833333
590    DEF FNXY(X) = (Define function here or in a
       subroutine)
750    END
```

POLYNOM.BAS

```
4000 POLYNOM: REM EVALUATES POLYNOMIAL
4010 REM N IS HIGHEST POWER OF X: K IS A DUMMY VARIABLE
4020 POL = COEF(N): K = N-1
4030 WHILE K >= 0
4040 POL = POL*X + COEF(K)
4050 K = K - 1
4060 WEND
4070 RETURN    : REM POL IS THE REQUIRED VALUE
```

ROUND.BAS

```
1000   ROUND : GOTO 1180 :REM ROUNDING SUBROUTINE FOR NM
1010   REM THE FOLLOWING LINES 1090 TO 1150 MUST BE
       INCLUDED
1020   REM (WITHOUT THE INITIAL "REM") SOMEWHERE IN THE
       MAIN
```

```
1030 REM PROGRAM OR BE RECALLED FROM DATA STORAGE.
1040 REM THE VALUES  L (DECIMAL PLACES) AND M
     (SIGNIFICANT
1050 REM  DIGITS) MUST BE SET EITHER POSITIVE OR ZERO AND
1060 REM C$ (CHOP) MUST BE EITHER "C" FOR ROUNDING DOWN
     OR "R"
1070 REM FOR NORMAL ROUNDING. MAX. NO. OF DIGITS IS 12 IN
     THIS
1080 REM LISTING, BUT IT MAY BE VARIED AS REQUIRED.
1090 REM RDTAB:  REM LOOK-UP TABLE SUBROUTINE FOR
     ROUNDING
1095 REM PROGRAM.
1100 REM DIM RD(12):REM ARRAY FOR HOLDING THE ROUNDING
     CONSTANT
1110 REM KO = 0.5
1120 REM FOR I = 1 TO 12
1130 REM RD(I) = KO * 10^(1-I)
1140 REM NEXT I
1150 REM RETURN
1160 REM FOR M SIGNIFICANT DECIMAL DIGITS
1170 REM OR L DECIMAL PLACES
1180 IF NM<0 THEN NM = -NM:FG$ = "-" ELSE FG$ = " ":IF NM
     = 0 THEN 1310
1190 IF M>0 AND L = 0 THEN 1210 ELSE 1200
1200 IF L>=0 AND M = 0 THEN 1330 ELSE 1320
1210 KK = 0 :IF NM = 1 THEN NM1 = 0.100000:KK = 1 :GOTO
     1250
1215 IF NM = 0 THEN KK = 0:NUM$ = " .0":LN = 1: GOTO 1270
1220 IF NM<1 THEN WHILE NM<.1:NM = NM*10:KK = KK-1
     :WEND: GOTO 1240
1230 IF NM>1 THEN WHILE NM>1:NM = NM*.1:KK = KK+1:
     WEND:GOTO 1240
1240 IF C$ = "C" THEN
     NM1 = NM ELSE NM1 = NM+RD(M+1)
1250 NUM$ = MID$(STR$(NM1),3)   :LN = LEN(NUM$)
1260 IF KK<0 THEN KK = -KK:GOTO 1290 ELSE 1270
1270 NUM1$ = FG$+LEFT$(NUM$,KK)+"."
1275 NUM1$ = NUM1$+MID$(NUM$,KK+1,M-KK)+STRING$(M-LN,"0")
1280 GOTO 1300
1290 NUM1$ = FG$+"."+STRING$(KK,"0")+LEFT$(NUM$,M)
1300 NM1 = VAL(NUM1$)
1310 RETURN
1320 PRINT "ERROR IN DATA - RESET L AND M":INPUT L,M:GOTO
```

```
     ROUND
1330  IF C$ = "C" THEN NM1 = NM ELSE  NM1 = NM+RD(L+1)
1340  NO$ = STR$(NM):NP = INSTR(NO$,"."):LN = LEN(NO$)
1345  IF INSTR(NO$,"E")<> 0 THEN PRINT NM,NMI,NO$:
      STOP
1350  IF NP = 0 THEN 1355 ELSE 1360
1355  PRINT"NUMBER IS INTEGER - EXAMINE PROGRAM":STOP
1360  IF LN-NP<L  THEN NM1 = NM:GOTO 1410  ELSE 1370
1370  NUM$ = STR$(NM1): LD = LN-LEN(NUM$)
1380  IF LD>0 THEN NUM$ = NUM$+STRING$(LD,"0")
1390  NUM1$ = LEFT$(NUM$,NP+L):MID$(NUM1$,1,1) = FG$
1400  NM1 = VAL(NUM1$)
1410  RETURN
```

## SIMP.BAS

This program is given in full, except for the definition of the function to be integrated and the necessary initial data, as an example.

```
100   REM PROGRAM SIMP.BAS (SIMPSON'S RULE)
120   FSP$ = "(Write function and date here)"
130   INPUT "RUN NUMBER   ", RNUM
140   DEFINT I-N:  NC = 18: REM NUMBER OF INTERVALS
160   (Define other constants here)
180   DIM YY(NC)
200   H = XRNGE/NC:  GOSUB FUNC : REM GIVES YY()
250   GOSUB SIMP
440   LPRINT "   AREA OF ";
460   LPRINT FSP$:LPRINT "          Run Number  ";
      RNUM:LPRINT
480   LPRINT TAB(5);"AREA =  ";TAB(15);SUMF
600   END

2000  SIMP:
2010  SUMF = 0
2020  FOR I = 0 TO NC
2050  IF I = 0 OR I = NC THEN SMP = 1:GOTO 2080 ELSE 2060
2060  IF I MOD 2 = 0 THEN SMP = 2 ELSE SMP = 4
2080  SUMF = SUMF+SMP*YY(I)
2100  NEXT I
2120  SUMF = SUMF*H/3
2140  RETURN
3000  FUNC:
```

```
3020  (Function to be integrated is defined here)
4000  RETURN
```

```
SIMPF.BAS
2000  SIMPF: REM INTEGRATION ROUTINE FOR FOURIER.BAS
2010  SUMF = 0 :NM = NC/2 :H = 2/3/NC
2020  FOR I = 0 TO NC
2030  XF = RNGE*(I-NM)/NC
2040  GOSUB FUNC
2050  IF I = 0 OR I = NC THEN SMP = 1:GOTO 2070 ELSE 2060
2060  IF I MOD 2 = 0 THEN SMP = 2 ELSE SMP = 4
2070  IF JF = 0 THEN FAC = 1 :GOTO 2080 ELSE 2075
2075  IF JF = 1 THEN FAC = COS(J*XF):GOTO 2080 ELSE FAC =
      SIN (J*XF)
2080  SUMF = SUMF+SMP*YY*FAC
2100  NEXT I
2120  SUMF = SUMF*H
2140  RETURN
```

SQROOT.BAS

```
10    SQROOT: REM SQUARE-ROOT ROUTINE
20    REM X0 IS START VALUE. Y IS NUMBER. H IS INCREMENT
30    PRINT "ENTER NUMBER "
40    INPUT Y : X0 = Y/2 : H = Y - X0*X0 : X1 = X0
50    ETA = 1E-8  : X = X0
60    WHILE H>ETA OR H< -ETA
70    X = (X*X+Y)/2/X : H = (Y-X1*X1)/2/X1 : X1 = H + X1
75    WEND
80    PRINT "SQR-RT OF " Y;" = ";X
90    END
```

STARTFP.BAS

```
3000  STARTFP: REM FINDS VALUES X0 AND X1 FOR FALSPOS
3010  X0 = XS : X1 = XS + H : REM XS + H IS STARTING
      RANGE
3020  Y0 = FNX(X0) : Y1 = FNX(X1) :I = 0
3030  REM FNX() IS DEFINED IN THE MAIN PROGRAM OR IN A
      SUBROUTINE
3040  WHILE I < N
```

```
3050    IF YO*Y1 < 0 THEN 3060 ELSE 3090
3060    PRINT "Do you want to try a smaller interval?"
3070    INPUT ANS
3080    IF ANS = "y" OR ANS = "Y" THEN H = H/2 : GOTO 3090
        ELSE RETURN
3090    Y0 = Y1 : X0 = X1
3100    X1 = X1 + H : Y1 = FNX(X1)
3110    IF Y0*Y1 > 0 AND I = N THEN 3120 ELSE I = I +1 :
        GOTO 3170
3120    PRINT "There is no root in this interval"
3130    PRINT "Do you want to continue?"
3140    INPUT ANS
3150    IF ANS = "y" OR ANS = "Y" THEN 3160 ELSE RETURN
3160    I = I + 1 : N = N + N
3170    WEND
3180    RETURN
```

# *WORKED SOLUTIONS TO EXERCISES*

---

## Solutions to Exercises 1.

### Question 1.

a). The appropriate limits are:

Diameter; lower limit – 12650 km
upper limit – 12750 km

Speed; lower limit – 750 km/hr
upper limit – 850 km/hr

b). If s = distance flown
t = time taken
v = speed
then t = s/v

| | | |
|---|---|---|
| therefore max. time | = 12750*π/4/750 | = 13.35 hr |
| and min. time | = 12650*π/4/850 | = 11.69 hr |
| the difference is | | 1.66 hr |

c). At £5000/hr the cost difference is £8300

**Question 2.** The Output Table produced on my computer is:

| *X* | *Y (direct)* | *1st Approx.* | *2nd Approx.* |
|---|---|---|---|
| 0 | 0 | 0 | 0 |
| 2 | 0 | 0.0013 | 0.0013 |
| 4 | 0.1192 | 0.0107 | 0.0107 |
| 6 | –0.1192 | 0.0360 | 0.0361 |
| 8 | 0.1192 | 0.0853 | 0.0855 |
| 10 | 0.3576 | 0.1667 | 0.1671 |
| 12 | 0.3576 | 0.2880 | 0.2889 |
| 14 | 0.4678 | 0.4573 | 0.4589 |
| 16 | 0.5960 | 0.6827 | 0.6854 |
| 18 | 1.0729 | 0.9720 | 0.9764 |

| | | | |
|---|---|---|---|
| 20 | 1.3113 | 1.3333 | 1.3400 |
| 22 | 1.9073 | 1.7747 | 1.7845 |
| 24 | 2.3842 | 2.3040 | 2.3179 |
| 26 | 2.9802 | 2.9293 | 2.9485 |
| 28 | 3.8147 | 3.6587 | 3.6844 |
| 30 | 4.5300 | 4.5000 | 4.5340 |
| 32 | 5.6028 | 5.4613 | 5.5053 |
| 34 | 6.4373 | 6.5507 | 6.6067 |
| 36 | 7.8678 | 7.7760 | 7.8465 |
| 38 | 9.4175 | 9.1453 | 9.2329 |
| 40 | 10.8480 | 10.6667 | 10.7742 |

This shows that the simple approximation is valid to three significant figures up to 10 μm and to better than 5% up to 50 μm.

## Solutions to Exercises 2.

1. a). Set t to 8

   b). Set t to 16. Introduce a new constant za.
   Set za to the (storage value of A) – 9
   Delete line 12
   Add an IF clause after line 11 as follows:

```
IF (fml > 9)
      num(i) = fml + za;
            ELSE
            num(i) = fml + ze;
ENDIF
```

2. Yes.

First prepare a look–up table in an array with the storage values of the numbers 0–9 and the letters A–F in consecutive locations. Then, in a corresponding array, put the binary codes of the numbers 0–15.

Move the mantissa into registers long enough to contain all the digits, then shift it two places to the right for octal and three places for hexadecimal adding zeros in front.

For each 1 in the exponent, from right to left, shift the mantissa the required number of places to the left into another register until the bias bit is reached. That is 1 for bit 0, two for bit 1, four for bit 2 and so on. Then insert the point after the last digit in the latter register. From the point, each three or four places left correspond to the required octal or hexadecimal integer digit respectively, and to the right to the corresponding fractional part. Using each group of digits as a mask, select the approriate number and print it out.

3. Because of rounding error, most machines give the answer zero to this question. Such effects must be considered when programs are written.

4. rrd is 0.4 for octal and 0.8 for hexadecimal (in decimal). It can always be expressed as 1/2 using the appropriate radix.

5. 175.989986 = 175 + 0.989986
176 – 175.989986 = 1 – 0.989986
Ans = 0.010014

6. In single precision, the first expression will overflow on most machines, whereas the second will nearly always give a valid answer. The correct value of both expressions is 30000.

7. a). Using the given data, the nominal mass of the block is 49.5 g. Assuming a length measuring error of 0.1 mm, the volume error is about 1%. The term ($1 - \rho a/\rho s$) occurs in all three expressions and may be considered constant. Call this constant b. Then the expressions become:

(a) $ms * b + \rho a * Vt$
(b) $ms * b /(1 - \rho a/\rho t)$
(c) $ms * (b + \rho a/\rho t)$

b). Formula a does not involve $\rho t$ and, as we are told that $Vt$ is constant, it will not be affected by variations in the density of the material. In this case it is worth making an effort to determine the value of the term $\rho a * Vt$ by a series of measurements in order to use formula *a*. In the event of a change in the block, the same formula can be used by substituting a new value for $Vt$ without the necessity of recalculating the errors.

c). Calculation shows that $\rho t$ can vary by 0.2%. The value of $\rho a/\rho t$ is 1.44E–3, so that the error in each of the formulas *b* and *c* is less than 2 parts per million which is less than the system resolution.

d). The mass of powder per block is 11 mg and therefore the errors in the formulas are negligible, but formula a is still to be preferred even at the expense of slightly greater complication because it involves a product and an addition as opposed to a straightforward product. (Each of the terms $b/(1 - \rho a/\rho t)$ and $b + \rho a/\rho t$ can be considered constant.)

# Solutions to Exercises 3

1. Log(10!) = 0+0.3+0.48+0.6+0.7+0.76+0.85+0.9+0.95+1
   = 6.54
   10! = 3.3E6 Correct value = 3.6288E6

2. Area of one cell = 0.8 * 0.05 * 0.05 = 2E–3 $m^2$
   Solar constant = 1.4 $kW/m^2$
   Power on 1 cell = 1.4k * 2 mil = 2.8 W
   Allow for atmosphere, say 2.5 W
   Power generated per cell = 0.5 W
   Number of cells required = 10000/0.5 = 20000 at equator
   At latitude 50° = 20000 / cos(50°) = 31000
   At the equator a frame about 5 m * 10 m would be required.

3. The drum of a concrete tanker is about 2 m diameter by 2 m long.

   a). The volume is approx. 6.5 cu. metre. The charge is then say, 2 cu. metre. The weight is 6 Tonne.

   b). Assume the thickness of concrete is 0.3 m
   Volume required is 75 cu. metre or 38 loads.

   c). It would be quite impracticable either to deliver or to work the concrete in one day. Such an amount is better mixed on site.

4. Distance from Los Angeles to London on the surface is about 10000 km. From the text the height of the satellite is 35000 km. Therefore the total travel distance is 80000 km. The velocity of light is 3E8 m/s.

   a). Travel time is 80 M / 300 M = 0.25 s

   b). The velocity of electro–magnetic information in a cable is 0.6 c (allowing for repeaters)
   Travel time is 10 M / 180 M = 0.05 s
   The difference is 0.2 s

5. The area of the billet is 0.09 $m^2$
   The wire area is 0.8 * 4E–6 = 3.2 μm
   Ratio = 90 mil / 3.2 μm = 30 k

   Now the maximum practicable winding speed for the wire is limited by the radial acceleration on the drum. For a drum speed of 750 R.P.M. with a diameter of 1 m the acceleration is 300 g and the corresponding wire speed is 40 m/s. In order to take up the through–put at this speed, 800 parallel output streams would be required with a first stage rolling speed of 1 m/s.

a). A continuous process is impracticable.

b). Not applicable

c). From the above argument, the maximum output velocity with a single die is about 25 m/s with a thicker bar. Then we have 0.8 * d * d = 0.09/25. This gives d = 0.06 m or 60 mm. The linear reduction ratio is 5 to 1 so that 2 stages would be necessary.

6. An orange is about 80 mm in diameter. The volume is then 4E–4 cu. metre. At a density of 2600 kg/cu metre the mass is 1 kg. The mass of gold is 10 μ kg and the volume is 0.5 n cu. m. The volume of one grain of gold is 0.8 f cu. m. Therefore the number of grains is 5/8 M = 600 000.

7. Allowing for losses the generator consumes about 200 W when all the electrical equipment is running. The engine supplies this extra power which is about 1% of the total. The fuel consumption must increase by roughly the same amount.

## Solutions to Exercises 4.

1. a). We have: $vo = -vi*A*R2/(R2+(A+1)*R1*(1+s*R2*C))$
Rearranging $vo = -vi*A/((A+1)*(1/(A+1)+R1*(1/R2+s*C))$
$= -vi*A/(A+1)*1/(s*C*R1+R1/R2+1/(A+1))$
Put $C*R1 = T1$, $C*R2 = T2$ and $u = -vi*A/((A+1)*s*T1)$
Then $vo = u*(1+1/(s*T2)+1/(s*T1*(A+1)))$
and, expanding the denominator
$vo = u*(1 - 1/(s*T2) - 1/(s*T1*(A+1)))$

This is similar to equation 4.19 with the addition of the term $1/(s*T2)$. If $T2 = (A+1)*T1$, the error term is twice as great. This occurs when $R2 = (A+1)*R1$.

b). If $R1$ = 1E6 Ohm and A = 1E6 then, for the error to be less than, or equal to, twice the initial error, $R2$ must be greater than 1E12 Ohm. As noted in the question, this value is difficult to achieve, so that it is safer to use a smaller value of $R1$, say 1E5 Ohm and either work with a smaller time constant $T1$ or use a larger capacitor. As the smaller time constant gives a much faster response, this is the preferred solution, providing the capacitor is accurate enough.

2. From 4.63 we have: $p = vm*vm/c$.
so that $pav = (1/c)*(v1*v1+v2*v2)/2$
and $vmeas = \sqrt{}((v1*v1+v2*v2)/2)$
but, $vav = (v1+v2)/2$
Put $Dv = v1-vav = vav-v2$
Therefore $vmeas = \sqrt{}(((vav+Dv)^{\wedge}2+(vav-Dv)^{\wedge}2)/2)$
$= \sqrt{}((2*vav^{\wedge}2 + 2*Dv^{\wedge}2)/2)$
$= \sqrt{}(1 + (Dv/vav)^{\wedge}2)$
$\approx vav*(1 + (1/2)*(Dv/vav)^{\wedge}2 + ..)$
If $v2 = v1/2$, then $Dv/vav = 1/6$
and the error becomes 1/72 or approx. 1.4%

3 a). Express equation 4.63 as $y = x*(1-2/x)/(x^{\wedge}2*(1-3/x)^{\wedge}2)$
which becomes $y = (1-2/x)/(x*(1-3/x))$
as $x$ tends to infinity, $y$ tends to $1/x$

b). Put $x = 1+h$; $y = (1+h-2)/(1+h-3)^{\wedge}2$
$y = (h-1)/(4*(1-h/2)^{\wedge}2)$
$\approx (h-1)*(1+h)/4$
$\approx -0.25*(1-h*h)$

Put $x = 2+h$; $y = h/(h-1)^{\wedge}2 \approx h*(1+2*h)$

Put $x = 3+\text{h}$; $y = (h+1)/h^{\wedge}2 = 1/h+1/h^{\wedge}2$

Note: The linear portion near $x = 2$ is very short and can only be used when $2*h$ is small enough to be neglected.

4 a). 20!/(9!*11!) = 20*19*18*17*16*15*14*13*12/9!
= 5*19*2*17*2*13*2
= 167960

b). = 1.67960E5 error = 0

Note: This value might be different for different machines.

c) Stirling = 1.67959E5 error ≈ 5ppm

d) Simp. Stir. = 1.70100E5 error ≈ 1.3%

5. a). The formula $x = (-b \pm (b^{\wedge}2 - 4*a*c))/(2*a)$ is very well–known, but should not be used when $(b^{\wedge}2-4*a*c)/b^{\wedge}2 < 0.1$ or when $4*a*c/b^{\wedge}2 < 0.1$. In the present case $4*a*c/b^{\wedge}2 = 5.6\text{E}-4$.
Rearranging and expanding by ihe binomial theorem, we get:

$x = -25*(1/2)*5.6E-4 + (1/4)*31.36E-8 -...$

for the negative value of $x$. The second term is too small to be considered in single precision and the value

$x = -0.007$ is found without a square root calculation.

b) The expression $x = 50 + 0.35/x$ is an example of an iterative solution in which an initial value is continually modified by substituting it back into the original expression. Given the right conditions the left hand side then converges to the correct answer. In this case the method works satisfactorily only for the positive root. The negative root is too close to zero and, unless the initial negative value is carefully chosen, the expression becomes unstable. Such equations must always be closely studied in order to find the appropriate roots. The technique is further examined in Chapter 6.

## Solutions to Exercises 5.

1. a). sinh^2($x$) – sin^2($x$) = (sinh($x$)+sin($x$))*(sinh($x$)–sin($x$))

Expanding these two factors, we get:

sinh($x$)+sin($x$) = 2*(x + x^5/5! +..... )

and sinh($x$)–sin($x$) = 2*($x$^3/3!+x^7/7!+... ).

The product $y$ is then:

$y$ = 4*($x$^4/3!+(x^8)*(1/7!+1/(5!*3!)+....)

$y$ = 2/3*($x$^4 + x^8/105 + .....)

For $x < 1$,the second term on the right hand side is always less than 1/100 and the higher terms are much smaller. Therefore the proposition is valid.

b). For $x = 0.3$ we have approximately

$y$ = 2/3 * 0.3^4

= 0.0054 with a fractional error of about 5E–5

c). 8 figure calculation gives 0.00540042

2. Let the area of the wire cross section be $A$, then the weight of the wire per unit length is $A*8000*g = w$

The tension $T$ is given by:

$T = A*8000*g*(z+c)$ from 5.14

but $z+c = c*\cosh(x/c)$ (catenary relation) so that when $x = 40$

$0.12 + c = c*(1 + x*x/(c*c*2) + ...)$

or $c = 40*40/(2*.12) = 6666.7$ m

and $T = A*8000*g*6666.7 = 5.23E8*A$ N

The stress is then 5.23E8 N/m$^2$ which is greater than the yield stress.

3. Because the computer algorithm is defined only at specific points, the given definition will leave indeterminate values between the points either side of zero. In most cases, the last value will be taken, thus producing a wave form with unequal positive and negative lengths.

The correct definition is:

$y = -1 \quad -\pi < x < 0$
$y = 0 \quad x = 0$
$y = 1 \quad 0 < x < \pi$

This makes certain that the cross–over point is defined.

4. In this question we assume that the volume of the glass does not change and that sections perpendicular to the axis remain perpendicular after drawing down. Then we have:

$(r\text{^}2-(r-t)\text{^}2)*l = (r1\text{^}2-(r1-t1)\text{^}2)*10*l$

and $r\text{^}2*l = r1\text{^}2*l*10$

| where | |
|---|---|
| $r$ | is the initial outside radius |
| $t$ | is the initial wall thickness |
| $r1$ | is the final outside radius |
| $t1$ | is the final wall thickness |
| $l$ | is the initial length |

simplifying gives:

$(r-t)$^2 = 10*$(r1-t1)$^2
or $(1-t/r)$^2 = $(1-t1/r1)$^2.

Expanding we have:

$-2*t/r \approx -2*ti/r1$

or $t1 = t/\sqrt{10}$.

5. The data set is taken from the equation:

$y = 100*(x$^3.5)*(1$-x)$^2.5, $0 < x < 0.9$

Using POLYFIT a ninth order curve was fitted giving the following coefficients:

| *Power of x* | *Coefficient* |
|---|---|
| 0 | 0 |
| 1 | 0.0163 |
| 2 | –0.8626 |
| 3 | 28.2243 |
| 4 | 63.6819 |
| 5 | –388.1741 |
| 6 | 596.3144 |
| 7 | –179.6120 |
| 8 | 240.3404 |
| 9 | –59.9070 |

The last two coefficients are not significant even though their values are quite large because $x$ is always less than 1. A curve and the derivative were plotted from these coefficients and the turning point estimated from the point at which the slope crossed the $x$ axis. The result obtained was $xm = 0.582$. From the curve the maximum value was estimated at $ym = 1.699$ by measurement.

Using FALSPOS and the slope equation, the corresponding figures were: $xm = 0.5833$ and $ym = 1.698$

The calculated values are: $xm = 0.5833$ and $ym = 1.6989$

The estimation of the $y$ value at $x = 1$ from the plotted points is very difficult and the only reasonable conclusion is that $y$ is less than 0.01 at $x = 1$. From the fitted

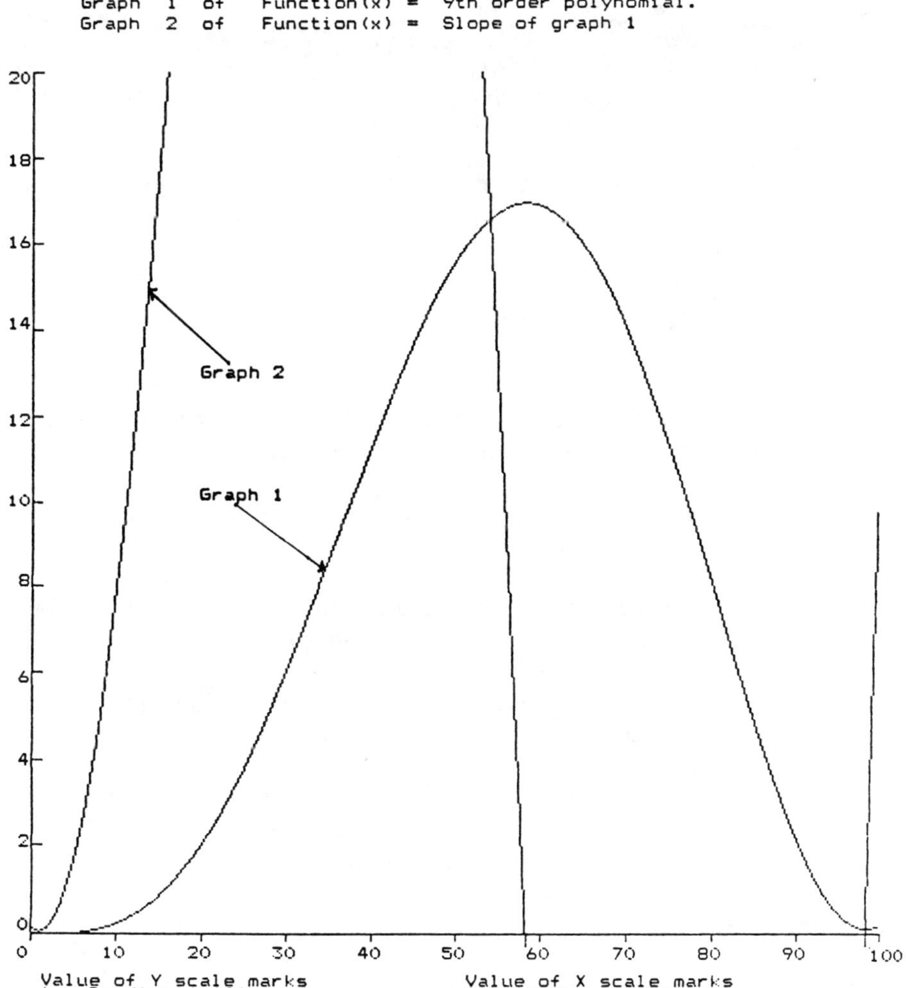


Fig. Ex. 5.5 Fit curve to given data set

polynomial, however, the value is 0.022 and the slope of the polynomial is positive instead of negative. This demonstrates that, although the fit is extremely close within the given range, it deteriorates very quickly outside it.

The curves are shown on the accompanying graphs.

6. a). Using FOURIER, the first 10 components of the saw–tooth are:

| *n* | *bn* |
|---|---|
| 0 | 0 |
| 1 | 0.637 |
| 2 | –0.318 |
| 3 | 0.213 |
| 4 | –0.160 |
| 5 | 0.128 |
| 6 | –0.106 |
| 7 | 0.091 |
| 8 | –0.080 |
| 9 | 0.071 |
| 10 | –0.064 |

b). Integration of the square of the straight line over the range with division by the range, gives the average power with a value of 1/3.

The mean of the square of a sinusoid over a complete cycle is 0.5 * (the square of the amplitude).

Half the square of the amplitude of the fundamental is 0.203 and therfore the power in this component is 61 % of the total.

c). Half the total sum of the squares of the harmonic amplitudes is 0.31 and there is thus about 6 % of the power unaccounted for by the 10 components.

A saw–tooth waveform is particularly rich in harmonics and therefore requires a wide bandwidth for its proper reproduction.

## Exercises 6.

1. a). $y = 1/(\sin(0.01)*\tan(0.01))$

Expanding the terms gives:
$y = 1/((0.01–0.01*0.01*0.01/6+....)*(0.01+0.01*0.01*0.01/3+...))$
$= 1/(0.0001*(1–0.01*0.01/6+...)*(1+0.01*0.01/3+...))$

$= 10000/(1+0.01*0.01*/6)$
$= 10000*(1-0.0000167)$
$= 9999.83$ to 6 figures

b). Similarly, expanding the b expression, we get:

$y = (1-0.01*0.01/2+..)/(0.01-0.01*0.01*0.01/6)/$
$(0.01-0.01*0.01*.001/6) = 10000*(1-0.0001/2)*(1+0.0001/3)$
$= 9999.83$ as before

c). The expansion this time is:

$y = (1-0.01*0.01/2+..)/(1-(1-0.01*0.01/2)*(1-0.01*0.01/2))$

giving:

$y = (1-.0001/2)/(1-(1-.0001))$

$= 9999.5$

The result for $c$ shows that, for the same precision, more terms are necessary in this case, and the calculation is more complex.

2. In order to reduce the effect of temperature as much as possible, it is sensible to adjust the system at the centre of the range, giving a maximum variation of 20 degrees.

Let $l$ be the range
$h$ be the range error
$b$ be the base line
$x$ be the base line error
$ang$ be the angle of the mirror
$alp$ be the angle error

Then we have:

$l+h = (b+x)*(\tan(ang+alp))$ and $l = b*\tan(ang)$

Assume that $x/b$ and $alp/ang$ are such that squares and higher terms may be neglected. Then expanding:

$l+h = (b+x)*(\tan(ang)+\tan(alp)/(1-\tan(ang)*\tan(alp))$

approximating:

$l+h = (b+x)*(\tan(ang)+alp)*(1+alp*\tan(ang)$
$= (b+x)*\tan(ang)+alp*b(1+\tan(ang)*\tan(ang))+x*\tan(ang)$ + product terms

a). Substituting for $l$ gives:

$h = alp*b + x*l/b + alp*l*l/b$

and rearranging we get a quadratic in $l$

$alp*l*l + x*l + alp*b*b - b*h = 0$ .

When the temperature is at its maximum,

$x = 5E{-}6*20*1 = 1E{-}4$

Now, in solving this equation after substituting the values for $b$,$x$,$h$ and $alp$, we find that the only important terms are b*h and alp*$l$*$l$. Therfore, we have:

$l*l = b*h/alp = 1E5$

and $l = 316$ m approx.

b). With $l = 2$ we get:

$h = 0.0002 + 0.0002 + 4*0.0002$

$= 0.0012$ m.

The baseline error is only important at short distances, where the system is very accurate, but the angle error makes the instrument useless at distances greater than about 300 m.

3. Equation 6.37 is:

$(P-Pv)\hat{}(3) - (P-Pv-g*ld*z)\hat{}(3) = 3*g*ld*C*t$

Let the diameter of the bubble at the surface, just before it bursts, be bd. Then from 6.33:

$P-Pv-g*ld*z = T/bd$. $z = 0.1$, $Bd = 5E{-}4$ m and $T = 0.055$ N/m
and $C = 2*g*ld*T*T*4/9/vc = 1.9682E4$.

Therefore $P\text{–}Pv = 1061.57$ and $g*ld*z = 951.57$

Putting $P\text{–}Pv = q$ and $g*ld*z = u$, 6.37 can be rewritten as:

$z*(q\verb|^|2+((q-u)*q)+(q-u)\verb|^|2) = 3*C*t$

or $t = z*(3*q\verb|^|2 - 3*u*q + u\verb|^|2)/(3*C)$

which gives, when the values are inserted:

$t$ = approx 2.19 s.

For a bubble of 1 mm diameter at the surface, the time is about 1.81 s.

4. a). Let $y\verb|^|3 = x$, then, if $y1$ is a first value, we have:

$(y1+h)\verb|^|3 = x$, where $h$ is the error.

Expanding and approximating:

$3*y1*y1*h = x - y1\verb|^|3$ or $h = (x-y1\verb|^|3)/(3*y1*y1)$

and $y2 = y1+h = y1+(x-y1\verb|^|3)/(3*y1*y1)$

$= (2*y1\verb|^|3+x)/(3*y1*y1)$

Start value is $x/3$ or $3*x$.

Similarly, for the fourth root:

$y2 = (3*y1\verb|^|4 +x)/(4*y1\verb|^|3)$

Start value is $x/4$ or $4*x$.

b). Cube root of 27.33 using the above method gives:

6.183, 4.362, 3.880, 3.192, 3.0221, 3.01114, 3.01212, 3.01217

27.33 = 3^3+0.33 = 27*(1+0.33/27)

Therefore cube root = 3*(1+0.33/27)^(1/3) = 3*(1+0.00407)
= 3.0122

Fourth root of 256.89 by iteration gives:

48.1615, 36.122, 27.092, 20.322, 15.25, 11.456, 8.635, 6.575, 5.158 4.335, 4.0407, 4.00398, 4.00348

and by binomial expansion 4*(1+0.89/256/4) = 4.00348.

Because of the poor starting point, the iterative methods were not so good as the expansions, but for values far from a known root they are much better.

6. a). Using SIMP with 18 intervals gives 1.77245 to 6 significant figures when the range is 0 to 4.5. Increasing the upper limit to 6 does not change the value at this precision.

b). Expanding $y$ = 2*exp(–$x$*$x$), we get:

$y$ = 2*(1–$x$*$x$+$x$^4/2–$x$^6/6+$x$^8/24–.....)

Integrating gives:

I = 2*($x$–$x$^3/3+$x$^5/10–$x$^7/42+$x$^9/216–....).

This expression converges very slowly and the terms are alternating. When $x$ is greater than 1, the terms increase in magnitude until (2*$n$+1)*n! becomes greater than $x$^(2*$n$+1). As the accompanying table shows, the successive initial terms are quite large, and the subtractions of two large numbers leads to such a loss of significance that the sum becomes meaningless unless the machine precision is adequate.

Table of the Integral of the Expansion of exp(–x*x)

| *n* | *term* | *sum* |
|---|---|---|
| 0 | 4.5 | 4.5 |
| 1 | –30.375 | –25.875 |
| 2 | 184.829 | 158.653 |
| 3 | –889.689 | –731.036 |
| 4 | 3503.15 | 2772.11 |
| 5 | –11608.1 | –8836.05 |
| 6 | 33150.2 | 24314.2 |
| 7 | –83112.4 | –58798.2 |

The terms increase until $n$=19, but the sum begins to decrease after the 18th. With single precision the error is already significant after the fourth term, and becomes worse as n increases up to about 40. Even with double precision, the error will be significant.

7. a). Using FALSPOS the first five roots of the equation are

| *Number* | *Root (rad)* |
|---|---|
| 1 | 4.49037 |
| 2 | 7.72521 |
| 3 | 10.90411 |
| 4 | 14.06618 |
| 5 | 17.22073 |

b). At the root of the equation, $y = 0$ and we aregiven that x is near $(2*n+1)*\pi/2$.

Put $(2*n+1)*\pi/2 = k$ and $z = k-x$.

Then, dividing by $\cos(x)/(x*x)$ and rearranging, we have:

$k-z = \tan(k-z)$, where $z/k < 1$.

From 6.02:

$k-z = 1/\tan(z) = 1/( +z\wedge 3/3+...)$

Neglecting higher powers we get:

$z*z - k*z + 1 = 0.$

Given that $z$ is small we may rewrite the equation as:

$z = z*z/k+1/k$. Then neglecting the first term we get

approximately, $x = k-1/k$

Therefore, with an of error less than 1%, we get:

| *Number* | *Root (rad)* |
|---|---|
| 1 | 4.50 |
| 2 | 7.73 |
| 3 | 10.91 |
| 4 | 14.07 |
| 5 | 17.22 |

## Solutions to Exercises 7.

1\. a). and b). The mean (*mn*) of the sum of *j* samples of a random variable tends to the Normal distribution with mean mn and standard deviation $sd/\sqrt{j}$. The table shows the normalised mean and standard deviation for 6 different values of *j* and the statistic 999**sd***sd*. The sample size in each case was 1000. If the value of the statistic is between 948 and 1052, there is less than a 5% chance that the distribution can be distinguished from the Normal by this method. For an additional test of the closeness of the distribution to the Normal, see question 3.

Table of sums of random numbers

| *Number in Sum (j)* | *Normalised Mean* | *Std. Dev.* | *999*sd*sd* |
|---|---|---|---|
| 5 | 0.0073 | 1.0172 | 1016 |
| 10 | –0.0191 | 1.0097 | 1009 |
| 15 | 0.0005 | 0.9871 | 986 |
| 20 | 0.0091 | 1.0049 | 1004 |
| 25 | –0.0051 | 1.0016 | 1001 |
| 30 | 0.0054 | 1.0063 | 1005 |

c). The results for this part of the question depend on the random generator of the machine, but they will generally be within the values given above even for sums as low as 5.

2\. a). This problem may be regarded as a binomial distribution with $P = 0.25$ and $n = 22$. Applying the rules of section 7.02, we find that $P*(1-P) = 0.18$ and $n*p = 5.5$. These values are marginal so that the normal approximation would be used only if the calculations were too complex to use the binomial equations direct.

b). The time taken is largely controlled by the number of red lights. The probability of a red light is 0.75, so that the average number is 22*0.75 = 16.5. Now, there is either a red light or there is not, so that to find the average speed for the journey we must take the mean of the speeds for 16 and 17 lights.

With 16 reds, the time is 20*60/40 + 16*.5 = 38 min.

With 17 reds, the time is 30 + 8.5 = 38.5 min.

The corresponding speeds are 31.58 and 31.17 respectively. The average speed is then 31.37 km/hr.

c). The probability of no red lights is (0.75)^0*(0.25)^22

giving: Prob(no reds) = 5.7E–14

This means that, with two journeys per day and 220 days per year, there would be a clear run once every 40 000 million years!

d). The probability of no greens is (0.25)^0*(0.75)^22

giving: Prob(no greens) = 0.0018 and a maximum time once every 15 months.

e). The limiting times are:

minimum = 30 min; maximum = 41 min and the average 38.25 min.

Strictly speaking the average time should not be calculated from the average speed because of the reciprocal relationship, but, in this instance, the difference is negligible.

f). Increasing the speed between lights to 60 km/hr would reduce the basic time to 20 minutes and therefore the average journey time would become 28.25 min.

3. a). In Pseudo–code a suitable program is

```
NORGEN.PSD

comment: successive random numbers from the machine
generator are assumed to be independent and can therefore
be used for RD1 and RD2. For histogram purposes a
convenient number of entries is 4212 = 18*234. To prevent
underflow on the logarithm, the number lo is set to a
conveniently small value, and entries less than lo are
discarded. This distorts the distribution slightly, but
the error is much smaller than that caused by other
factors. endcomment

    CONST: INT ni,lo;    REAL pi2,k;
    VAR  : INT i;        REAL y,z,anorm;
```

```
BEGIN
  ni = 4211; k = -2;
  pi2 = 2*3.14159265;
  OPEN (output), #1, "norm.dat"; comment: opens file.
                                 endcomment
    DO (i = 0; i = ni; i = i+1;)
      y = RANDOM; comment: first random number.
      endcomment
        IF (y < lo)
          CONTINUE
        ELSE
         z = RANDOM; comment: second random number.
                              endcomment
        ENDIF
      anorm = SQRT(k*log(y))*cos(pi2*z)
      WRITEDSK, #1, anorm
    CONTINUE
  CLOSE #1
 END
```

b). There are three possible methods for this part of the question. The first is to sort all the numbers in ascending order and then determine the fraction between 0 and 1. The second method is to fit a curve to a histogram of the values and integrate the polynomial using SIMP. Thirdly a polynomial can be fitted to the sorted numbers.

| | |
|---|---|
| The first method gives | $P = 0.3389$ |
| The second | $P = 0.3421$ |
| The third | $P = 0.3431$ |
| From tables | $P = 0.3413$. |

c). As question 1 shows, the simple tests for Normality do not give a precise answer, For a Normal distribution, the first four moments are 0, 1, 0, and 3. All the distributions in question 1 are close to Normal and for four–figure precision the sum of 20 uniformly–distributed random numbers will give a satisfactory result for many purposes. For special cases it is adviseable to test the distribution using standard methods.

4. a). Fitting a curve to the product of the normal and the chisquare(4) curve gives the following coefficients using units of £1000 on the $x$ axis:

| *Order* | *Coefficient* |
|---|---|
| 0 | –0.0859089 |
| 1 | 1.29539 |
| 2 | –1.23500 |
| 3 | 0.538124 |
| 4 | –0.135223 |
| 5 | 0.021143 |
| 6 | –2.09182E–3 |
| 7 | 1.27435E–4 |
| 8 | –4.36327E–6 |
| 9 | 6.42478E–8 |

b). The mean of the above distribution is £1944

c). The most probable daily turnover is £1024

d). The probability of taking more than £10000 in one day is about 0.0021 or about 1 in 470

e). The probability of a daily turnover near the mode is of the order of 0.4, while the probability of a turnover near the mean is only about 0.2. Therefore if the remaining time for the estimate is less than one month, the most accurate prediction would be obtained using the mode. However, the central value selected necessarily depends on the purpose for which the information is required.

Note 1. Owing to the alternate signs in the coefficients, it is VERY important that, to reduce rounding error, the coefficients should be given to at least six decimal digits.

Note 2. Due to random variations in the data, the histogram values at low probabilities fluctuate wildly. The fitted curve attempts to follow these fluctuations and, as a result, the graph is not reliable at x values greater than 8000.

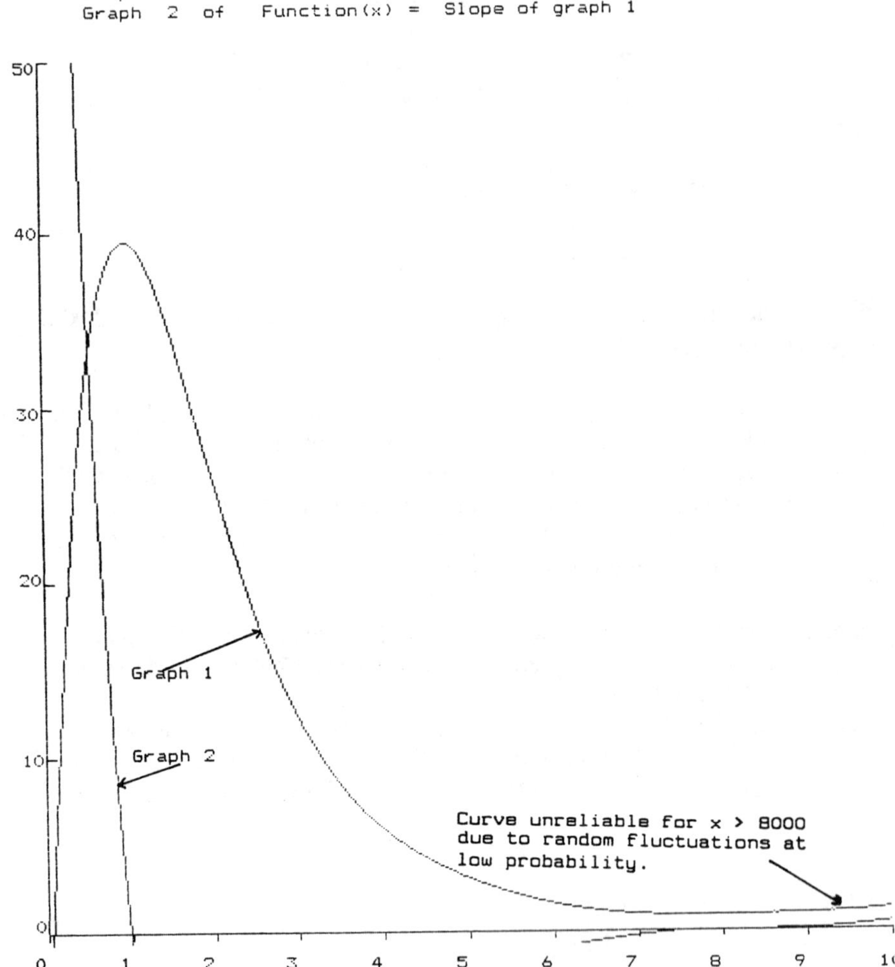


**Fig. Ex 7.4** Probability of shop turnover

# INDEX